MATH BY EXAMPLE

MATH BY EXAMPLE

THE WONDERFUL WORLD OF MATHEMATICS

Marina Grinman

iUniverse, Inc.
New York Lincoln Shanghai

Math by Example
The Wonderful World of Mathematics

iUniverse, Inc.

For information address:
iUniverse, Inc.
2021 Pine Lake Road, Suite 100
Lincoln, NE 68512
www.iuniverse.com

ISBN: 0-595-32185-2

Printed in the United States of America

Contents

Preface

While teaching math courses, I was giving away word problems handouts and my students encouraged me to create a book containing all of my own word problems. This is how this book was born. My main goal was not to write another SAT preparation manual, but rather the good supplement for one. After a month of solving five to ten problems a day, student develops more advanced approach to SAT. My students were able to boost their SAT score by hundred or more points just by solving word problems.

The book may also prove to be useful to the math teachers: today's textbooks do not contain enough diverse word problems.

Thanks

First of all, a big "Thank you" to my students for their support during all these years.

Also, a big thanks to my family:
my husband Alex, a manager for New York City Transit, and a good mathematician himself for his encouragement and valuable feedback;

Hy Farber, who did most of the proofreading while being allergic to math;

my son Lev for his positive attitude.

I am also grateful to excellent young artist Lauren Arguelles for creating the cover of this book.

Problems

1) A motorboat travels at the rate of 12.8 miles per hour. The river current flows at the rate of 0.7 miles per hour. Find the rate of the boat going against the current. Find the rate of a boat going with the current.

2) Perform the indicated operations.

$$5 \times [50 - 8 \times (8.3 + 9.2) \div 3.5]$$

3) Find the fraction that names rational number as each decimal.

a) $0.\overline{5}$ b) $-0.\overline{07}$ c) $0.\overline{65}$ d) $-0.\overline{414}$

4) Maria had $220.30. She spent 6 times less money in the supermarket than in the department store. How much money did she spend in the department store if she still has $10.30 left?

5) Evaluate.

$$[-3.57 + (-2.86)] + [-4.3 + (-2.27)]$$

6) Carlos has a number in mind. If this number is increased 11 times and the result decreased by 2.75, then the resulting number will be 85.25. What is Carlos's original number?

7) Solve and check.

$$3\frac{2}{5} \div x = 6\frac{4}{5} \div 1\frac{1}{3}$$

8) Tim has a number in mind. If this number is increased by 9.2 and the result increased 11 times, then the resulting number will be 110. What is Tim's original number?

9) Perform the indicated operations.
$$51 - (3.75 \div 3 + 86.45 \div 24.7) \times 2.4$$

10) A shop had 9 gallons of O. J. If 22% of the juice was sold on a first day and 12% on the next day, how many more gallons of juice were sold on the first day?

11) Find the solution set of the equation.
$$\frac{3(x-12)}{2} = x + 1 - \frac{3(2x-4)}{4}$$

12) A deli department of a supermarket has 8 pounds of cold cuts. If 43% is used to make sandwiches and 33% sold to the customers, how many fewer pounds of cold cuts did the customer buy?

13) Evaluate.

$$7.35 \times (8 - 2.4) + (9 - 3.2) \times 6.45$$

14) There are 1,720 trees in a certain orchard; 25% are cherry trees and 15% apple trees. How many cherry trees are in the orchard? How many apple trees?

15) Simplify an expression by combining like terms.

$$3.7y + 2.5y + 1.6y + 4.8y + 0.6 + 0.8$$

16) A concession stand at the football game sold 5,000 candy bars. 12% of all bars sold were small bars and 13% were large bars. How many small and large candy bars were sold?

17) Solve and check.

$$4\frac{2}{5} \div x = 8\frac{4}{5} \div 2\frac{1}{2}$$

18) Nord Music Store took a survey of 1,232 people. 1.2 times more people said that they liked rap better than rock. How many people preferred rap?

19) Perform the indicated operations.

$$11 + (56.28 \div 6 - 108.9 \div 19.8) \times 2.5$$

20) Milk contains 3.9% fat. How many pounds of fat are in 800 pounds of the same milk?

21) Find the solution set of the equation.

$$\frac{2(5x+2)}{3} - 1 = 3 + 2x - \frac{2(1-11x)}{3}$$

22) A tip of 15% is standard for good service in a restaurant. If Bill wants to leave a tip of about 15% on a lunch check of $23.50, how much should he leave?

23) Solve for the variable in each equation and check.

   a) $8.05 \div x = 2.3$
   b) $y \times 2.7 = 10.881$

24) On a recent tour a popular singer traveled 840 miles, visiting three cities. The distance between Richmond and Knoxville was 3 times the distance from Raleigh to Richmond and the distance between Raleigh and Knoxville was twice a distance from Raleigh to Richmond. What is the distance between Richmond and Knoxville?

25) Perform the indicated operations.

$$111-[(0.9744 \div 0.24+10.2) \times 2.5-2.65]$$

26) 250 black, gray, and red horses participated in a Belmont Stakes race; 30 of them were black, 0.7 of the rest were gray. How many red horses participated in the race?

27) Solve and check.

$$7\frac{1}{3} \div 2\frac{1}{2} = 3\frac{2}{3} \div y$$

28) A hitchhiker plans a 40-mile trip. On the first day he covers 40% of the whole distance and on the second day he covers 30%. How many miles does he have to hike on the third day to finish the trip?

29) Solve for x and check.

$$16\frac{1}{2} - 9.12 \div x = 4\frac{1}{2}$$

30) A drug store carries two different–sized bottles of contact lens solution. There is a total of 500 bottles in the store, and 225 of them are size A. What percent are size B bottles?

31) Evaluate.

$$(20 - 5.163) \div 3.7 - 16.006 \div (3 + 2.3)$$

32) David walked for 1.8 kilometers when his dog started to run after him and reached him in 3 minutes. How fast did David walk if the dog's speed was 0.7 kilometers per minute?

33) Find the solution set of the equation.

$$\frac{2(x+10)}{5} = -\frac{2(2x-5)}{3}$$

34) A dog ran at a speed of 19 kilometers per hour to overtake his owner when the distance between them was 1.8 kilometers. How fast was dog's owner jogging if the dog reached him in 0.2 hours?

35) Simplify an expression by combining like terms.

$$4.5x + 1.9x + 0.7x + 1.6x + 5.7x + 0.9$$

36) Two trains leave a station at the same rate of speed. One train travels for 9 hours and the other for 5 hours. How fast were the trains moving if the second one traveled 290 miles less than the first one?

37) Evaluate.

$$[(10.5 + 8.06) \div 5.8 - 2.2] \times 3.87$$

38) A car and a truck leave towns A and B at the same time moving towards each other. A distance between towns is 280 miles. A car moves with the speed of 80 miles per hour while a truck speed rate is 75% of that of the car. In how many hours will the two meet?

39) Find the solution set of the equation.

$$y \div 320 - 31 = 79$$

40) Two cyclists start moving towards each other at the same time. The first bicycle is moving 20 miles per hour and the other one is 25% slower. How long will it take them to meet if a distance between their starting points is 70 miles?

41) Find the numerical value of each expression.

      a) $|x|-|y|$. Use $x = -64.1$, $y = 52.8$

      b) $|x|+|y|$. Use $x = -54.5$, $y = 7.6$

42) To make cranberry jam one needs 3 pounds of sugar for each 2 pounds of cranberries. How much sugar is needed to make 10 pounds of jam if processing reduces the total weight of the final product 1.5 times?

43) Transform the expression into a polynomial.

      $(0.4b - 0.5c)^2$

44) A lesson lasts for 45 minutes, 9 of which were used for mental exercise. What percent is this?

45) Simplify first and write the number that the numerical expression represents.

$$\frac{48^2 + 2 \bullet 48 \bullet 18 + 18^2}{48^2 - 18^2}$$

46) Joe sails upstream on a motorboat for 3 hours, and then he rafts back. The boat's speed is 24 miles per hour, and the current is moving 3 miles per hour. How long does the return trip take?

47) Solve for x.

$$(387.75 - x) \div 2.5 = 168.75 - 41.25$$

48) Mary rafted 75 miles along with the current for 25 hours, and then she took a motorboat back. The speed rate of the boat is 28 miles per hour. How much time did she need for the return trip?

49) Simplify.

$$(\frac{1}{2}a + \frac{1}{3}b) - (\frac{5}{2}a - \frac{2}{3}b) + (a + b)$$

50) Sam and John live in towns 36 kilometers apart. They start walking toward each other at the same time and meet in 4 hours. Sam walks at the speed of 4 kilometers per hour. Find John's speed rate.

51) Solve for y.

$$280.44 \div (y - 42.6) = 121.62 - 51.51$$

52) Cindy sold 76 candies. She sold twice as many white chocolate hearts as milk chocolate hearts and 8 more milk chocolate assortments than milk chocolate hearts. How many of each candy type did she sell?

53) Find the solution set of the equation:

$$13,200 \div x - 2,180 = 20$$

54) In one baseball season, fans in the Oceanside school district bought a total of 171 bags of peanuts, candy bars and potato chips. The number of peanuts purchased was 2 times less than that of potato chips; the number of candy bars purchased was 23 more than that of peanuts. How many candy bars did the Oceanside baseball fans purchase this season?

55) Perform the indicated operations.

$$200 - [(9.08 - 2.6828 \div 0.38) \times 8.5 + 0.83]$$

56) The popular toy, "walking wiry coil", was selling with a great success. In three days one store sold 9,100 of them. The first day twice as many toys were sold as the next day. On the third day a store sold 700 more than it did on the second day. How many of these were sold on the first day?

57) Solve and check.

$$6\frac{1}{2} \div 3\frac{3}{4} = 3\frac{1}{4} \div a$$

58) On a trip to Florida, a car traveled 4 hours at the rate of 62.5 miles per hour, then 2.5 hours at the rate of 53.4 miles per hour, and finally 1.5 hours at the rate of 82.2 miles per hour. Find the car's average speed.

59) Write the number represented by numerical expression.

$$[44.96 + 28.84 \div (13.7 - 10.9)] \div 1.8$$

60) A European gourmet store mixed three types of tea—Orange pekoe, Ceylon, and Russian black—to create a special blend. They used 300 grams of Orange pekoe tea that costs 96 cents per 100 grams, 450 grams of Ceylon tea costing $ 1.04 per 100 grams, and 250 grams of Russian black tea costing 76 cents per 100 grams. How much did 100 grams of the mix cost? (Round the answer to whole cents).

61) Transform the expression into a polynomial.

$$\left(\frac{2}{3}x^3 - \frac{3}{4}\right) \times \left(\frac{2}{3}x^3 + \frac{3}{4}\right)$$

62) Kurt Harris charged $380 for a three-day project. On a first day he billed the company 2.4 times more money than for the second day, which left a balance of $106.30. How much money did Kurt get on a first day?

63) Solve for x and check.

$$(3\frac{2}{3} x - 1.2) - 4.4 = 1$$

64) A real estate agent helps a family to choose an apartment that meets their needs. Find the square footage of the master bedroom if the total area of the apartment is 400.64 square feet, the master bedroom 3 times larger than the kitchen, and the rest of the apartment measures 240 square feet.

65) Write the number represented by numerical expression.

$$102.816 \div (3.2 \times 6.3) + 3.84$$

66) There are 40 players on the Sailors team. There are 12 players on the field at one time. What percent is this?

67) Transform the expression into a polynomial.

$$\left(\frac{1}{4} a^3 - \frac{4}{5} \right)^2$$

68) A Boy Scout team planted 60 trees on the outskirts of their town. 3 of these trees have not survived the rough winter. What percent of all trees survived the winter?

69) Find the solution set of the equation.
$$\frac{2(x-4)}{3} + \frac{x+3}{2} = \frac{2x-3}{3}$$

70) Ann paid $12.20 for 4 pounds of candy and 3 pounds of cookies. If she bought 2 pounds of the same candy and 3 pounds of the same cookies, she would end up paying $8.20. How much did a pound of cookies cost?

71) Solve the equation for x.
$$\frac{b}{x} = \frac{b}{a}$$

72) During railroad construction in Texas, two tunnels were build. How long would it take for the 750 meter-long train to pass entirely thru a 9 kilometer tunnel if it takes 21 minutes for this train to pass thru a 15 kilometer-long tunnel?

$$(1 \text{ kilometer} = 1000 \text{ meters})$$

73) Perform the indicated operations.

$$(2\frac{1}{3} \times 4\frac{1}{2} - 2.5 \div 2\frac{1}{2}) \div 4\frac{1}{2}$$

74) Working together, Tom and Jerry can paint a 58 square feet wall in 2 hours. While working alone, Jerry can paint 5 square feet per hour more than Tom. How many square feet per hour can Tom paint?

75) Solve the equation for x.

$$\frac{x}{a} = \frac{ab}{b}$$

76) Amy travels 79.2 miles to get to the tourist camp by bicycle. The first 48.6 miles of her trip is along a country road on which she averages 12.15 miles per hour, and rests for 2.5 hours. The second part is on a highway where she averages 15.3 miles per hour. What is the total time of the trip?

77) Evaluate.

$$[-0.251 + (-0.37)] + [-0.2 + (-0.152)]$$

78) Natasha is making a 32.4-kilometer hiking trip. She travels first 4.5 hours at 5.2 kilometers per hour, and then rests for 1.6 hours; the last part of her trip is a bumpy road on which she walks at 2.5 kilometers per hour. How much time did Natasha spend for the whole trip?

79) Transform the expression into a polynomial.

$$(-4ab - 5a^2)^2$$

80) Peter is going to visit his friend. If he travels at 3.4 miles per hour for 2 hours, and then at 9.1 miles per hour for 1 hour, what is the average (arithmetic mean) speed of his entire trip?

81) Simplify first and write the number that the numerical expression represents.

$$\frac{85^2 - 17^2}{85^2 + 2 \bullet 85 \bullet 17 + 17^2}$$

82) An alloy contains 16.4 pounds of copper and 3.6 pounds of nickel. What is the percentage of copper in the alloy?

83) Write the number represented by numerical expression.

$$[27.59 - (3.52 + 2.18) \times 1.5] \div 2.72$$

84) At a company meeting workers decided to vacation in July and August. 51 workers took time off in July, leaving 83% of workers to vacation in August. How many people took vacation in August?

85) Solve the equation for x.

$$\frac{a-b}{b} = \frac{a-b}{x}$$

86) Melinda spent $30 on a blouse, leaving her with 85% of her original amount to buy groceries. How much money did Melinda spend on groceries?

87) Factor the polynomial completely.

$$125 + 30a + 6a^2 + 25a$$

88) 29 geese and 46 ducks were staying on Birds Island during the winter. 15 of these birds were white-feathered. What percent of the birds were white-feathered?

89) Prove the identity within the set of real numbers.

$$\frac{a+b}{x-y} \times \frac{x-y}{(a-b)(a+b)} = \frac{1}{a-b}$$

90) David Reznick wishes to fertilize his two small lawns—35 and 40 square feet each. A walking path occupies 9 square feet of the lawns. What percent of the area should not be fertilized?

91) Transform the expression into a polynomial.

$$(-3b^2 - 2ab)^2$$

92) Sam has twice as much money as his brother Jim. If Sam spends $580, two of them together will have $2,360 left. How much money did each brother have originally?

93) Factor completely.
$$-3a^2 - 12ab - 12b^2$$

94) USSR used to be the largest country in the world. The northern part of this country is covered with woods, swamps, and meadows. Woods cover 480,000 square miles; swamps cover 370,000 less square miles than woods, and meadows cover 5 times more than swamps. What is the area of meadows?

95) Evaluate.
$$(5.05 \div \frac{1}{5} - 2.8 \times \frac{5}{7}) \times 3 - 1.6 \times 0.01875$$

96) 15 cubic centimeters of copper and 10 cubic centimeters of zinc are melted together. How much will 1 cubic centimeter of resulting alloy weigh, if 1 cm³ of copper weighs 8.9 grams and 1 cm³ of zinc weighs 7.1 grams? (Round to the nearest tenth of a gram).

97) Solve for the variable in the equation and check.

$$16x - (4x - 5) = 15$$

98) To make a box, Michael uses a piece of cardboard measuring 40 square inches. He cuts off 3.2 square inches. What percent of a cardboard is left?

99) Factor the polynomial completely.

$$m^3 - 12m^2 + 4m - 48$$

100) Tamika borrowed a 240-page book from the library. On Saturday she has read 7.5% of the whole book, and on Sunday – 12 pages more than on Saturday. How many pages are left to read?

101) Prove the identity within the set of real numbers.

$$\frac{a+b}{a(a-b)} \div \frac{a+b}{b(a-b)} = \frac{b}{a}$$

102) A pet store has a stock of 260 pounds of a bird food. During the first month they sell 8.5% of the total. The second month they sell 30 pounds more than the first month. How much of the bird food still is in stock?

103) Transform the expression into a polynomial.

$$(0.2x^2 + 5xy)^2$$

104) Peter's mother is baking pies for a charity event at the school. She separates flour into three piles. The ratio of their weights is 2:3:5. Find the weight of each pile if the third pile weighs 6.3 ounces more than the first one.

105) Solve for the variable in the equation and check.

$$64x - (3 - 8x) = 87$$

106) Marta wants to make a milk shake by using milk, syrup, and ice cream in the ratio 15:2:3. How many grams of each ingredient should she use, if she takes 240 more grams of milk than the ice cream?

107) Factor the polynomial completely.

$$x^3 - 3x^2y + 4y^2x - 12y^3$$

108) Dan has two empty fish tanks. The table below shows their measurements in inches.

Fish tanks	Length	Width	Height
First tank	52	35	38
Second tank	7	52	50

If Dan fills the second tank to its capacity and then empties it into the first one, how high would the water level be in the first fish tank?

109) Perform the expression into a polynomial.

$$(4xy + 0.5y^2)^2$$

110) An average human inhales 0.55 liters of air, making 15 inhales in one minute. What is the total volume of air inhaled in one hour?

111) Factor completely.

$$-2a^2 + 8ab - 8b^2$$

112) A hiker and a bicyclist start moving towards each other from 40 miles apart. The bicyclist's rate of speed is 4 times greater than that of a hiker. Find the speed rate of each if they are going to meet in 2.5 hours.

113) Solve for the variable in the equation and check.
$$-5(x-3)-5(x-1)=-20$$

114) Two trains started from towns A and B toward each other. The distance between towns is 210 miles. The speed rate of the train going from town A is 5 miles per hour more than the speed of the other train. If they meet in two hours, what was the rate of each train?

115) Write an expression in factored form.
$$a^2x^2 - bx^2 + a^2x - bx$$

116) Basketball shoes regularly priced at $106 are now discounted by 15%. How much do they cost?

117) Factor the polynomial completely.
$$c + 3c^2d + 3cd^2 + d$$

118) There were 42 books on one library bookshelf and 34 on another bookshelf. Some books were removed from both shelves. The amount of books removed from the first shelf was equal to the number of books remaining on the second shelf after the books were removed. As a result 12 books were left on the first shelf. How many books were removed from the second shelf?

119) Solve an equation.
$$(13.4 - y) \times 4.3 - 20.05 = 78.05 + 6.7y$$

120) A basketball team scored 42 points in the first quarter. In the second quarter the team scored 3 points less than it did the third quarter. If the team scored 82 points in the first two quarters, what is the number scored in the third quarter?

121) Solve for the variable in the equation and check.

$$(2x - 3) + (2x + 3) = 12$$

122) A bookstore has 240 copies of math textbooks. 15% are Algebra textbooks, which is 2/3 of Trigonometry books. If the rest of the math textbooks are Geometry books, then how many copies of Geometry textbooks are in the store?

123) Simplify.

$$\frac{a - 5}{(a - 3)(a + 3)} \times \frac{a + 3}{a - 5}$$

124) In one year New Yorkers went on 150,000 trips by bus, train, and motor home. 14% traveled by bus, which is 7/8 of all who traveled by train. How many traveled by motor home?

125) Factor the expression.
$$(x - y)(x + y) + x + y$$

126) New immigrants to Israel decided to plant an orchard. Out of 350 trees planted, 2/5 were apple trees, which is 70% of all planted apricot trees. The rest of all trees planted were cherry trees. How many cherry trees were planted?

127) Factor completely.
   a) $2a - 6 + 3a - 9$
   b) $5a + 10 - 10a - 20$

128) A department store sold 450 shirts of different fabric pattern. Solid color shirts were 2/5 of all shirts sold, which is 90% of all striped shirts. The remaining shirts were plaid. How many plaid shirts were sold in the department store?

129) Find the product.

$$\frac{1}{2xy} \times \frac{4x^2 y^3}{5z} \times \frac{10z^2}{3x^3}$$

130) Long Island Railroad train covers a distance between stations in 6 hours at the rate of 68 miles per hour. How long would it take a bicycle to cover 1/8 of the train route, if it goes 17 miles per hour?

131) Factor completely.

$$(2a + b) - (4a^2 - b^2)$$

132) An American Airlines plane covers the distance between Toronto and Miami in 3 hours going 640 miles per hour. If Mr. Smith takes a train to cover 1/6 of the whole distance that running at the speed rate of 80 miles per hour, how much time would it take?

133) Find the numeric value of the expression.

$$5a^2 - 5ax - 7a + 7x.\ \text{Use}\ x = -3, a = 4$$

134) The average (arithmetic mean) of two distinct positive numbers is 13.05 and one is 4 times less than the other one. Find both numbers.

135) Find the product.

$$\left(\frac{5a}{7b}\right)^2 \bullet \frac{14b^2}{25a^3}$$

136) The average (arithmetic mean) of two distinct positive numbers is 12.32 and one is 3 times greater than the other one. Find both numbers.

137) Simplify.

$$\frac{b-4}{b+3} \div \frac{(b-4)^2}{b^2-9}$$

138) A bakery uses 26 pounds of a special flour mixture that contains corn, wheat, and rye. If a bag mixture contains 2 times more corn than wheat, and 8 pounds of rye, how many pounds of corn and wheat are in the flour mixture?

139) Simplify the expression.

$$(x+y)^2 - (x^2 - 2xy + y^2)$$

140) A boy skates 3 times faster than his sister and they are 96 meters apart. How fast is he skating, if they meet in 8 seconds?

141) Factor completely.

        a) $x^2 + 8x + 16$

        b) $9m^2 + 6mn + n^2$

142) In a list of 3 integers, the first is 60, the second is 80% of the first one, and the third one is 50% of he sum of the first and the second. Find an average (arithmetic mean) of all three integers.

143) Solve an equation.

$$(16.2 - x) \times 3.2 - 50.08 = -8.12 - 5.1\,x$$

144) The treasurer of a club invested $12,120 of the club's savings into stocks, bonds, and treasury notes. 30% of the money was invested in stocks. How much did the treasurer invest in bonds and how much in treasury notes, if he spent $6,244 more on bonds than on treasury notes?

145) Find the product.

$$\left(\frac{3a^2}{2b}\right)^3 \cdot \frac{16b^3}{21a^4}$$

146) A spider is moving up a tree at a rate of 6 centimeters per second. A caterpillar positioned 60 centimeters below the spider starts moving down the tree. At what rate is the caterpillar traveling if in 5 seconds the distance between the two insects will be 100 centimeters?

147) Simplify.

$$\frac{a^2 - 49}{(a+b)^2} \times \frac{a+b}{a-7}$$

148) Karen and David are riding bicycles towards each other at the same rate. The distance between them is now 40 miles, but in 3 hours the distance will be 50 miles. Find their rates of travel.

149) A 234-meter bridge was built over a river. It has 5 spans; 4 out of 5 spans are the same length and the fifth one is 14 meters longer than the other 4. What is the length of each span?

150) Factor completely.

$$\text{a) } 1 + 2c + c^2$$
$$\text{b) } 81 - 18x + x^2$$

151) Mr. Taylor forgot to pay a $200 bill. After the late charge was applied, he ended up paying $260. By what percent was the original payment increased?

152) Write an expression in factored form.

$$ax - bx + ay - by - ax + bx$$

153) A motorcycle leaves town A for town B, which is 250 miles away. At the same time, a motor scooter started a trip from town B to town A. In 2 hours, the distance between the vehicles will be 30 miles. If the rate of the motorcycle is 10 miles per hour greater than that of the motor scooter, what are the rates of travel of both vehicles?

154) Find the product.

$$\left(\frac{ab}{cd} \right)^2 \times acd$$

155) A train left A for B, which is 261 miles away. After 2 hours it increased its rate by 12 miles per hour and arrived at B 3 hours following it change of speed. What was the speed of the train at the beginning of the trip?

156) Factor completely.

$$a) \ 100 - 20a + a^2$$
$$b) \ a^2 + 10ab + 25b^2$$

157) Suppose you go to Smith's Sporting Goods to buy equipment for in-line skating and you want to spend $300. After you finish shopping, the cashier tells you that the total is $375. By what percent have you exceeded the amount planned?

158) Find the product.

$$abc^2 \times \left(\frac{ab}{cd} \right)^2$$

159) A motorboat travels 59.2 miles downstream and 87.5 miles against the current. If the river flows at a rate of 2.3 miles per hour and the motorboat moves at a rate of 27.3 miles per hour, how much time did the whole trip take?

160) Find the numeric value of the expression.

$$m^2 - mn - 3m + 3n. \text{ Use } m = 0.5, n = 0.25$$

161) A shopper spent $8.10 on vegetables consisting of potatoes and string beans. If potatoes cost $1.00 per pound and string beans cost $.70 per pound and the shopper bought 3 pounds more potatoes than beans, how much of each vegetable was purchased?

162) Add.

$$7a^2 + 2b^2 - (6a^2 + b^2)$$

163) Karin walked at a steady rate for 3 hours and then rode a bicycle for another 4 hours, also at a constant rate. The total distance traveled was 62 kilometers. What was her walking speed, if she rode the bicycle 5 kilometers per hour faster than she walked?

164) Perform the indicated operations.

$$\frac{8a^2b}{9c} \times \frac{36c^3}{5a^3b}$$

165) A train leaves the station at 75 miles per hour. Another train is coming from the opposite direction at 90 miles per hour. What would be the distance between the two trains 5 hours after they pass each other? What would be the distance between them half an hour before they meet?

166) Factor the expression.

$$x + x^2y + xy^2 + y$$

167) A boat sailed for 2 hours at the rate of 5.2 miles per hour, then for 3 hours at the rate of 6.4 miles per hour, and finally for 5 hours at the rate of 5.5 miles per hour. Find the average speed of the boat.

168) Simplify.

$$(0.3a - 1.2b) + (a - b) - (1.3a - 0.2b)$$

169) Two cars move toward each other, one at a rate of 60 miles per hour and the other at 80 miles per hour. What would be the distance between them 3 hours after they pass each other? What would be the distance between them ½ an hour before they meet?

170) Simplify.

$$\frac{a-1}{2a+1} \div \frac{a-1}{(2a-1)(2a+1)}$$

171) A traveler hiked for 2 hours at the rate of 6.3 kilometers per hour, rode a bicycle at the rate of 12.6 kilometers per hour for 3 hours, and then sailed at the rate of 9.9 kilometers per hour for 4 hours. Find the traveler's average speed.

172) Find the solution set of the equation.

$$2\frac{1}{3} \times k = 4\frac{1}{2} \times 1\frac{1}{9}$$

173) Maria bought 3 pairs of pants and a blouse for a total of $100. How much did each item cost, if the blouse cost $8 more than the pair of pants?

174) Multiply.

$$(1\frac{4}{7}a^3x^3 - 2\frac{3}{4}a^2x^3 - 11ax^4) \times (-2\frac{6}{11}ax^6)$$

175) The class president asked his classmates to help fight pollution. A group of students organized a letter writing campaign. They sent 119 letters to the president, their senator, and their governor. The students mailed 4 letters more to the president than to the senator, but 3 letters less to the governor, than to the president. How many letters did they mail to each official?

176) Find the numerical value of the expression.

$$(x+1)(x+2)+(x+3)(x+4). \text{ Use } x=-0.4$$

177) A mail carrier picked up two bags of mail. The first bag was 3 times heavier than the second one. If the carrier moves 20 pounds of mail from the heavier to the lighter bag, then their weights would be even. What was the initial weight of each bag?

178) Find the numeric value of the expression.

$$a^2 + ab - 5a - 5b. \text{ Use } a=6.6, b=0.4$$

179) The length of line segment AB is 2 inches more than the length of line segment CD. If the length of segment AB is increased by 10 inches and the length of segment CD is increased 3 times its original length, the lengths of both segments become equal. Find the original length of AB.

180) Simplify.

$$\left(\frac{a}{a+1}+1\right) \div \left(1-\frac{a}{a+1}\right)$$

181) It takes 1.8 hours to get to town B from town A by bus. Traveling by car would take 0.8 hours to complete the same trip. A bus moves 50 miles per hour slower than the car. Find the rate at which the bus is traveling.

182) Add.

$$4a^2 + 2a + 1 - (1 + 2a - 4a^2)$$

183) Mr. and Mrs. Carr own a small orange grove. On the first day Mr. Carr picked 1.2 times more bushels of navel oranges than Mrs. Carr picked. The second day he picked 1.4 bushels of oranges while his wife picked 1.9 bushels. Considering that both of them picked equal amount of bushels in two days, how many of them did Mr. Carr pick on the first day?

184) Multiply.

$$(\frac{2}{3}a^2b^4 + \frac{1}{2}a^3b) \times \frac{3}{2}ab^3$$

185) A newspaper club collects bundles of newspapers. The students use two wagons to take the bundles to the recycling plant. The first wagon carries 0.6 tones more than the second one. If they load 0.2 times more paper on the first wagon and 0.4 times more on the second one then they would both carry the same load. How much paper does each one carry?

186) Simplify.

$$(11p^3 - 2p^2) - (p^3 - p^2) + (-5p^2 - 3p^3)$$

187) Two trucks are traveling at a steady rate. There are 100 gallons of fuel in the first truck's tank and 70 gallons of fuel in the second truck's tank. After the 20-hour ride the amount of fuel left in the first truck's tank is equal to the amount of fuel left in the second truck's tank. If the first truck used 1.5 times more fuel than the second truck, how much fuel each truck uses per hour?

188) Find the solution set of the equation.

$$8\frac{1}{2} \times m = 3\frac{2}{3} \times 1\frac{1}{11}$$

189) Pat and Angel are coworkers. They are training to run the New York City marathon. By their plan both Pat and Angel have to run the same total distance in two days. The first day Pat ran 3 miles more than Angel. The next day Pat ran 1.5 times more miles than she had during the first day while Angel ran twice as much as he had the first day. How many miles did each one of them run the first day?

190) Factor the expression.

$$10x^2 + 10xy + 5x + 5y$$

191) The American Kennel club registered 68 different breeds over a 4-month period. During the first two months the same number of breeds were registered. In the third month two times more breeds were registered than during the first month. In the fourth month, 3 more breeds than during the second month. How many breeds were registered each month?

192) Simplify.

$$\left(\frac{a}{a-1}+1\right)\div\left(1-\frac{2a}{1-a}\right)$$

193) The larger of two numbers is 4.5 times greater than the other. If 54 is subtracted from the larger number and 72 is added to the smaller number then both results will be equal. What are the numbers?

194) Multiply.

a) $(\frac{1}{2}a+3b)(\frac{1}{2}a-3b)$

b) $(0.3-m)(m+0.3)$

c) $(\frac{1}{3}a-2b)(\frac{1}{3}a+2b)$

195) A bicycle leaves point A traveling 60 miles per hour slower than a motorcycle leaving point A and traveling in the same direction. It takes the bicycle 1.5 hours to get to point B, while the motorcycle reaches the destination in 0.25 hours. Find the rate of the motorcycle and the distance between point A and B.

196) Multiply.

$$(-2\frac{4}{9}b^6y + 2\frac{1}{5}b^3y^2 - 11by^5) \times (-2\frac{1}{22}b^4y^5)$$

197) Sue and Sam have equal amounts of money. If Sue buys a book and Sam buys a CD that costs 1.4 times less than Sue's book, then Sue would have $2 left while Sam would have $4 left. What is the price of the book? The CD? How much money did each friend start with?

198) Find the solution set of the equation.

$$y \div \frac{4}{5} = 3\frac{1}{8} \div 1\frac{1}{4}$$

199) Boy scouts planned to recycle 72 aluminum cans, but instead 90 cans were recycled. What percent of the aluminum cans was recycled compared to that planned?

200) Simplify the expression.

$$(1-a)(1+a)-(1-a)^2$$

201) Members of the sixth-grade class planned to raise $640 for the wildlife preserve, but instead raised $720. By what percent did the six-graders exceed the amount of money they planned to raise?

202) Simplify the expression.

$$5x^3 \div x - (2x)^2 + x^4 \div (2x^2)$$

203) Shoe Universe offers Monika a temporary job during her spring break. The store manager gives her a choice as to how she wants to be paid, but she must decide before she starts working.
Plan 1: $140 a day, working 8 hours a day, 5 days a week.
Plan 2: $120 a day, working 6 hours a day for 5 days, plus 17% of a week's wage to work 6 hours on Saturday.
What would be Monica's weekly rate under each plan?

204) Solve for x.

$$x \div \frac{3}{14} = 3\frac{1}{9} \div \frac{4}{9}$$

205) To make a new dried fruit mix a manufacturer uses 150 pounds of fresh cherries. How many pounds of dried cherries will the factory make if processing reduces the total weight of fresh cherries by 80%?

206) Simplify.

a) $a^7 \times a^3 \times a^{-5}$

b) $x \times x^8 \times x^{-1}$

207) A police boat cruising along the river at the rate of 40 miles per hour follows a motorboat traveling at the rate of 25 miles per hour. The distance between them is now 30 miles. What will be the distance between the two vessels in 1 hour?

208) Factor the expression.

$$5ax + ay - 10y - 50x$$

209) A World of Best Clothing store ordered new stock of coats, dresses, pants and blouses. The order consisted of 45% coats, 30% dresses, 15% pants, and 400 blouses. How many new pieces of clothing did the store order?

210) Factor the expression.

$$2a^2 - 3ab + 14ac - 21bc$$

211) Mr. Jensen is planning a movie festival for the Markus movie theater. Of all the movies he ordered 30% were westerns, 40% were comedies, 10% were mysteries, and 60 were science-fiction adventures. How many movies did Mr. Jensen order?

212) Find the numeric value of the expression.

$$a^2 - ab - 2a + 2b. \text{ Use } a = \frac{7}{20}, b = 0.15$$

213) When John drove by the farmer's market in the morning he noticed that the price of apples was the same as the price of pears. When he drove back in the afternoon the price of apples was reduced by 20 cents a pound while the price of pears was reduced by only 10 cents. John bought 6 pounds of apples and 5 pounds of pears and paid as much money for apples as he paid for pears. What was the price of one pound of apples in the morning?

214) Solve for x.

$$3\frac{2}{5} \div x = 6\frac{4}{5} \div 1\frac{1}{3}$$

215) Farmer Joe is packing mini-pumpkins for the Halloween. He put the same number of pumpkins into each basket and each box. To make baskets lighter for his kids to carry he removed 10 pumpkins from each basket and then he put 20 more pumpkins into each box. Now 14 baskets contain as many pumpkins as 10 boxes. How many pumpkins did Joe put in each box originally?

216) Solve and check.

$$\frac{14}{7} = \frac{2-x}{5}$$

217) Two rolls contain the same length of tape. When 10 feet of tape were cut off the first roll and 40 feet off the second roll, then the first roll had twice as much tape as the second one. How much tape was on each roll originally?

218) Multiply.

$$(\frac{1}{2}a^3b^2 - \frac{3}{4}ab^4) \times \frac{4}{3}a^3b$$

219) The Smith family is going to have Saturday breakfast. When father went to the bagel store he bought 6 plain bagels, 4 poppy seed bagels, and cream cheese. Considering that all bagels cost the same, cream cheese cost $1.30, and he paid a total of $7.30, how much did each bagel cost?

220) Add.

$$\frac{3y}{x^2 - y^2} + \frac{2x}{y^2 - x^2}$$

221) Simplify each expression.

a) $\dfrac{a^6 \times a^4}{a^5}$

b) $\dfrac{3^3 \times 3^2}{3^5}$

222) A small bag of potato chips and two chocolate-chip cookies have a total of 700 calories. If the chips have 200 calories less than 1 cookie, how many calories are in the bag of chips?

223) Solve and check.

$$\frac{x}{3} + \frac{x}{5} = 8$$

224) For many years, immigrants to the United States arrived at Ellis Island in New York. During 1902, approximately 740,000 people came from Germany, Russia, and Ireland. 380,000 more people came from Russia than from Germany. Considering that the same number of immigrants came from Germany as from Ireland, how many people arrived from Russia?

225) Factor the expression.

$$4xz^2 + 3x^2z - 15x^2z - 20xz^2$$

226) Karla has paid $8.30 to develop 3 rolls of film in three different places. A local drug store charged $0.30 more for a roll of film, than a mail-order film developer. A local developer charged 2 times more than a drug store. How much money did the mail-order developer charge?

227) Simplify and evaluate.

$$(12 \times 5 - 8 \times 5^2 + 4 \times 5^3) \div (4 \times 5^2)$$

228) There are 430 students at West High School. The number of freshmen is 3 times more than the number of sophomores. The number of sophomores is 35 less than the number of juniors. How many freshmen, sophomores, and juniors are at West High School?

229) Multiply.

$$(a^2 - ab + b^2) \bullet 3ab^3$$

230) The sum of four consecutive integers is equal to 2. Find all four numbers.

231) Simplify the following expression.

$$\frac{m^2}{m-n} - \frac{m^2}{n-m} + \frac{n^2}{m-n}$$

232) The sum of five consecutive integers is equal to –10. Find all five numbers.

233) Simplify.

$$\frac{1-a}{1+b} \times \frac{b+b^2}{a-a^2}$$

234) When 500 cafeteria customers were asked about their dessert preferences, 40% chose cheese-cake. Chocolate cake was preferred by 12% fewer people than those who chose cheesecake. How many people liked cheesecake, and how many liked chocolate cake?

235) Solve an equation.

$$\frac{x}{5} = \frac{6+x}{3}$$

236) There are 360 girls at New Middle School. If the number of boys is 52% of the total number of students, how many students attend New Middle School?

237) Simplify the expression.
$$(12x^3 - 8x^2) \div 4x - 4x(3x + 0.25)$$

238) Ann traveled during her summer break. After the trip Ann decided to calculate how many miles she traveled. She found out that the railroad trip was 120 miles longer than the boat ride. If she traveled 4 times more miles by train and 8 times more miles by boat than she actually traveled, the distance would be 1,200 miles. What was the actual distance traveled by boat?

239) Find a product.
$$(-0.4x^5 y^6 z^2) \times (-1.2xyz^3)$$

240) Helen and Terry each opened checking accounts and deposited equal amounts of money. If Helen deposited $80 more and Terry withdrew $20, then Helen would have 3 times more money in her checking account than Terry. How much money did the girls deposited originally?

241) Perform the indicated operations.

$$\frac{(x-y)(x+y)}{6xy} \times \frac{12x^2 y}{x+y}$$

242) The length of Pete's stride is 12 centimeters greater than Mike's. Actually 4 of Pete's strides are shorter than of 6 of Mike's by 54 centimeters. How long is the stride length of each boy?

243) Simplify the expression.

$$(3x^4 + \frac{1}{3}x^2) \div x - x^3 \div (3x^2) - 3x^3$$

244) Two boats were cruising along the coastline of Maine. At the beginning twice more people were on the first boat than on the second boat. When at the first stop 98 people disembarked the first boat and 16 disembarked the second boat, the number of passengers became the same on both boats. How many people were on board each boat at the beginning?

245) Solve for y.

$$\frac{y}{3} + \frac{y}{4} = 14$$

246) Right before Thanksgiving Day Donovan's supermarket stocked 3 times more turkeys than Melnik's supermarket. After Donovan's sold 960 turkeys and Melnik's received a delivery of 240 more, both supermarkets had the same number of turkeys. How many turkeys were stocked at Donovan's supermarket and at Melnik's supermarket originally?

247) Factor each expression.

a) $7(y-3) - a(3-y)$

b) $b^2(a-1) - c(1-a)$

248) Martha opened checking and savings accounts with twice as much money in savings. When she transferred $140 from the savings account to the checking, balances on both accounts became equal. How much money did she deposit in the checking account originally?

249) Simplify the following expression.

$$\frac{a}{a^2 - 1} - \frac{1}{1 - a^2}$$

250) A certain number is multiplied by 2 and 50 added to the product. The sum is multiplied by 5, and 200 is subtracted from the new product. Then the result is divided by 10. The result is 30. Find the number.

251) Solve for x.

$$(-2)^3 \times x + (0.4)^2 = (-1)^9 - (1 - 2x)$$

252) While working on the Global Volunteer's Day workers of Glenn Brokerage Firm were separated into two teams. The first team was cleaning up, and the second team was painting the benches and playgrounds in the park along East River. In the morning there were 3 times more people on the painting team than on the cleaning team. Later on, when 12 more workers of the Firm joined the cleaning team both teams had 76 people combined. How many people are in the cleaning team now?

253) Factor each expression.

a) $6(a-2)+a(2-a)$

b) $a^2(m-2)+b(2-m)$

254) A group of children went on a day trip to a local farm. A farmer gave them 123 oranges and 82 apples that the children split equally. How many children were on the trip? How many oranges and apples did each child get?

255) Simplify each expression.

a) $\dfrac{(2x-3)(2x+3)}{2x+3}$

b) $\dfrac{(b-2)^2}{b^2-4}$

256) A bus depot received an order to dispatch a number of buses to drive the workers of the Green Company to the baseball game and picnic. 424 workers went to the baseball game and 477 went to the picnic. How many buses were used if each bus had the same number of seats and all of them were filled to capacity?

257) Determine, for each pair, which is the greater number.

a) 0.6 or $0.\overline{6}$

b) 0.31 or $0.\overline{31}$

c) $-\sqrt{7}$ or -7

d) 0.6 or $\sqrt{0.6}$

258) A jar of preserves costs $0.85 and a jar of jam $0.60. Nelly paid a total of $6.90 for 9 jars. How many of each did she buy?

259) Evaluate each expression.

a) $\dfrac{(-0.2)^4}{(0.1)^5}$

b) $\dfrac{(0.3)^3}{(-0.1)^4}$

c) $\dfrac{(3.2)^2}{(1.6)^2}$

d) $\dfrac{(2.6)^2}{(1.3)^2}$

260) Write a product as a polynomial.

$$(-2\frac{4}{9}b^6y + 2\frac{1}{5}b^3y^2 - 11by^5) \times (-2\frac{1}{22}b^4y^5)$$

261) A magazine costs $0.60 and a book costs $1.40. Roy paid a total of $28.60 for 25 publications. How many of each did Roy buy?

262) Simplify the expression.

$$6x^4 \div x - 5x^5 \div x^2 + (2x)^3$$

263) Basketball shoes and one pair of socks cost $145; 50 pairs of socks cost $59 more than a pair of shoes. How much does each pair of socks cost?

264) Factor each polynomial completely.

      a) $ac - 3bd + ad - 3bc$

      b) $5ay - 3bx + ax - 15by$

265) Out of 80 freshmen in King's College, 21.25% study French. Out of 90 juniors 20% study Spanish. Which language is more popular among students?

266) Perform operations.

$$(\frac{1}{2}a + \frac{1}{3}b) - (\frac{5}{2}a - \frac{2}{3}b) + (a + b)$$

267) Larry has a total of $5.30 in nickels, dimes, and quarters. He has as many nickels as dimes, and two more quarters than dimes. How many dimes does he have?

268) Find a product.

$$(-2.5n^4m^5k^2)\times(3nm^2k^5)$$

269) Two cruise ships, *Beautiful* and *Wonderful,* leave the port A at the same time headed for port B. *Beautiful* needs 4 days for the round trip, while *Wonderful* needs 6 days. How many days would it take for both ships to meet again at the port A?

270) Solve the equation.

$$(1.2)^2 - (0.1)^2 \times (20 - 200x) = (1.4)^2$$

271) Three cruise-liners start their journeys from port A at the same time. The first one returns to the port every 15 days, the second–every 20 days, and the third–every 24 days. How many days would it take for them to leave port A together again if they start the next cruise immediately after they return?

272) Evaluate each expression.

a) $\dfrac{2^5 \times 2^3}{2^4}$

b) $\dfrac{3^{11} \times 9}{3^{12}}$

c) $\dfrac{3^4 \times 3^5}{3^8}$

d) $\dfrac{2^6 \times 16}{2^3}$

273) To make a fruit jelly Mrs. Kerry takes 2.5 pounds of apple, 2 pounds of pears, and 0.5 pounds of cherries. What is the percentage of each fruit in the mixture?

274) Simplify each expression.

a) $\dfrac{a^2 - 16}{a - 4}$

b) $\dfrac{4 - a^2}{a + 2}$

275) If a height of a closet is 7.5 feet, the width is 40% of its height, and the length 150% of its width, what is the volume of the closet?

276) Find a product.

$$(-1\frac{1}{3}x^2y^3z)\times(-1\frac{1}{2}xy^2z^3)$$

277) A motorboat going downstream takes 4 hours to travel a certain distance. It takes the motorboat 6 hours to make the return trip against the current. If the river flows at the rate of 2.5 miles per hour, find the rate of the motorboat in still water and the distance traveled one way.

278) Simplify.

$$(0.3a-1.2b)+(a-b)-(1.3a-0.2b)$$

279) The length of a rectangular solid is 21 inches, the width is 4/7 of its length, and the height is 2/3 of its width. Find the volume of the solid.

280) In each case factor completely.

$$a) \ ac + bc - 2ad - 2bd$$
$$b) \ 2bx - 3ay - 6by + ax$$

281) An airplane started from Kennedy Airport at a speed of 600 miles per hour. In 0.5 hours another plane started in the same direction at a speed of 750 miles per hour. How long would it take for the second plane to be 225 miles ahead of the first plane?

282) Write a product as a polynomial.

$$(1\frac{4}{7}a^3x^3 - 2\frac{3}{4}a^2x^3 - 11ax^4)(-2\frac{6}{11}ax^6)$$

283) A bus started from town A at the rate of 60 miles per hour. In 0.5 hours a car started from the same point going in the same direction at 75 miles per hour. How long would it take for the car to be 45 miles ahead of the bus?

284) Write the number that each numerical expression represents.

a) $2\times10^5 + 3\times10^4 + 5\times10^3 + 1\times10^2 + 2\times10 + 1$

b) $3\times10^6 + 5\times10^5 + 3\times10^4 + 2\times10^3 + 3\times10 + 7$

c) $1\times10^5 + 1\times10^3 + 1$

285) Mr. Black walks to his workplace located 6 kilometers from his home. He makes 100 steps, 0.80 meters long each, in 1 minute. When does he have to leave his house if he is planning to come to work 10 minutes before 8:30 A.M, considering that 1 kilometer = 1000 meters?

286) Perform operations.

$$11p^3 - 2p^2 - (p^3 - p^2) + (-5p^2 - 3p^3)$$

287) On a route between New York and Albany, 135 miles apart, a bus makes 5 intermediate stops, 2 minutes long, and spends 45 minutes at the Albany terminal. The bus travels at 45 miles per hour. At what time will the bus be back in New York if it left at 10:45 in the morning, and it takes the same amount of time each way?

288) Simplify.

$$\frac{a-b}{ab} - \frac{a-c}{ac}$$

289) Joan invested a certain amount of money earning 3% interest per year. Find the amount she invested if at the end of a year she had $412.

290) Perform the indicated operations.

$$\frac{4ab-4b^2}{a^2+ab} \times \frac{a+b}{4b^2}$$

291) David is cycling from the town he lives in at the rate of 15 miles per hour. In 12 minutes he overtakes his friend Jerry running in the same direction. How fast does Jerry run if the initial distance between them was 1.8 miles?

292) Simplify the expression.

$$(2m^2n^3 - 3m^2n^2 + 4n^2m^3) \times \frac{1}{12}m^2n^2$$

293) The garment factory is planning to make 180 wedding gowns. Using 32 inches of fabric less per dress, the factory can make 192 dresses. How much fabric does one dress take?

294) Evaluate each expression.

a) $\left(\dfrac{3}{5}\right)^4 \times \dfrac{5^3}{3^2}$

b) $\dfrac{7^5}{5^7} \times \left(\dfrac{5}{7}\right)^6$

c) $\left(\dfrac{2}{3}\right)^3 \times \left(\dfrac{3}{2}\right)^5$

d) $\left(\dfrac{3}{4}\right)^6 \times \left(\dfrac{4}{3}\right)^8$

295) During a typing competition, Angela finished her assignment in 2.9 hours. If she has typed 12 words more per minute, the assignment would be finished in 2.8 hours. How many words per minute did Angela type?

296) Factor each polynomial completely and check the result.

$\quad$ a) $16\,ab^2 - 5b^2c - 10c^3 + 32ac^2$

$\quad$ b) $-28ac + 35c^2 - 10cx + 8ax$

297) Amy takes 6 hours to make 16 paper toys for the Girl Scout charity event, while Julie makes 24 toys in 15 hours. Who works faster?

298) Perform the indicated operations.

$$\frac{a^2 + 4a}{a - 4} \div \frac{a + 4}{a^2 - 4a}$$

299) A fisherman caught 2 eels, each 20 inches long, 5 pikes–each 16 inches long, and 1 snapper–12 inches long. Find the average length of fish caught.

300) Write a product as a polynomial.

$$(\frac{2}{3}a^2b^4 + \frac{1}{2}a^3b) \times (\frac{3}{2}ab^3)$$

301) A college school of agriculture is growing corn and sunflowers. Its cornfield produces 60 bushels from each acre. Its sunflower field produces 15 bushels from each acre. How big will the crop be, if there are 40 acres of the cornfield, while the sunflower field is 1.75 times more?

302) Simplify.

$$5x^2 + 5x^3 + (x^3 - x^2) - (-2x^3 + 4x^2)$$

303) Mrs. Garcia picked out three different fabrics to make pillows for her living room. She picked 324 square inches of dark-colored fabric, which was 4 times more than the amount of light-colored fabric, and 256 square inches of linen. How many square inches has she bought in total?

304) Perform the indicated operations.

$$\frac{5a^2b + 5ab^2}{a - b} \div \frac{10ab}{3a - 3b}$$

305) Ms. Green sells fabric. There are two rolls of high quality wool in her store. The first roll has twice as much fabric as a second one. After she cuts off 15 feet of fabric from the second roll and 45 feet from the first one, both rolls of fabric will be equal in length. How many feet of wool did each roll have at the beginning?

306) Prove an identity.

$$(n^2 - 2n)(n^2 + 2n) = n^4 - 4n^2$$

307) A narrow pipe can fill a water tank in 10 hours. A wider pipe takes 4 hours to complete the same job. Which pipe will pour more water in a water tank–a wide pipe working for 3 hours or a narrow pipe working for 7 hours?

308) Find a product.

$$(2\frac{1}{4}a^2b^5c^3)\times(-3\frac{1}{3}a^3b^2c^4)$$

309) A rope that is 3 feet long is cut into 7 equal pieces. A rope that is 4 feet long is cut into 10 equal pieces. Which one produces longer pieces?

310) Multiply.

$$(5c-4y)\times(-8c-2x+6y)$$

311) Alex can paint the room in 14 hours. It takes Sam 8 hours to do the same job. Who would accomplish more work – Alex working for 7 hours or Sam working for 5 hours?

312) Write an expression in factored form.

$$xy^2 - by^2 - ax + ab + y^2 - a$$

313) It takes a bus 8 hours to cover a certain distance, while a car takes 6 hours to travel the same distance. Which vehicle would go farther–a bus in 5 hours or a car in 4 hours?

314) Evaluate.

a) $139 \times 15 + 18 \times 139 + 15 \times 261 + 18 \times 261$
b) $14.7 \times 13 - 2 \times 14.7 + 13 \times 5.3 - 2 \times 5.3$

315) A car departed from point A at noon traveling at the rate of 70 miles per hour. Three hours earlier a truck left from the same point going in the same direction. At what time will the car pass the truck, if the speed of the truck was $\frac{4}{7}$ of the car?

316) Add.

$$\frac{1}{a^2} + \frac{1}{ab} + \frac{1}{b^2}$$

317) Nelly plans to spend $59 to buy cashews, peanuts, and almonds for her company's Fourth of July Party. She spends 5 times more on peanuts than on cashews and $10 more on almonds than on cashews. What is the cost of each type of nut?

318) Multiply.

$$(4b - c)(-5b + 3c - 4y)$$

319) When getting ready for a trip, Debra spent $65 to buy perfume, deodorant, and suntan lotion. The deodorant cost $15 more than the suntan lotion. Together the deodorant and suntan lotion together cost $1 more than the perfume. How much did Debra pay for each item?

320) Simplify an expression.

$$12.5x^2 + y^2 - (8x^2 - 5y^2 - [-10x^2 + (5.5x^2 - 6y^2)])$$

321) Dan completes a certain job in 6 hours, George takes 8 hours to complete the same job. If they work together, what part of the job would they complete in 1 hour?

322) Perform the indicated operations.

$$\frac{a^2 + 2a}{(a-1)(a+1)} \times \frac{(a-1)}{(a+2)}$$

323) Amy and Ann start walking toward each other from points A and B. Amy takes 6 hours to walk from A to B, while it takes Ann 5 hours from B to A. What part of the whole distance will they cover if they walk for 1 hour only?

324) Write a product as a polynomial.

$$(\frac{1}{2}a^3b^2 - \frac{3}{4}ab^4) \times (\frac{4}{3}a^3b)$$

325) One pipe can fill a tank in 4 hours. A second pipe takes 6 hours to complete the same job. If they work together, what part of the tank will still be empty after 1 hour?

326) Simplify.

$$\left(\frac{b-a}{1+ab}\right) \div \left(\frac{a(b-a)}{1+ab}\right)$$

327) A standard tip for service in a restaurant is 15%. If Gregory paid $230 for the dinner with friends, tip included, how much did the food cost?

328) Simplify the expression.

$$(8m^3 - 7m^2 n + 1) \times \frac{1}{8} mn$$

329) Dillon bought a pair of shoes for $55, including 10% sales tax. What was the price of the shoes before tax?

330) Evaluate.

a) $25 \times 114 - 31 \times 13 - 25 \times 83$

b) $3\frac{1}{3} \times 4\frac{1}{5} + 4.2 \times \frac{2}{3} + 3\frac{1}{3} \times 2\frac{4}{5} + 2.8 \times \frac{2}{3}$

331) Mel, an interior designer, bought set of chairs to furnish an office for his client. If he put 3 chairs in each room, he would need 3 more chairs; if he places 2 chairs in each room, there would be 3 extra chairs. How many chairs did Mel buy?

332) Find the numerical value of the expression.

$$(m-5)(m-1)-(m+2)(m-5). \text{ Use } m=-2\frac{3}{5}$$

333) A catering company bought a number of small cakes for a party. If they place 3 cakes on each table, there will be 6 cakes left, if they place 4 cakes on the each table, they will need 2 more cakes. How many cakes did the catering company buy for the occasion?

334) Divide and express the quotient in lowest term.

a) $8abc \div (-4a)$

b) $(-10pq) \div (6q)$

335) A helicopter delivered a 3-month supply to the polar station. The cargo was dropped from the helicopter and landed in 3 seconds. What was the height the cargo was dropped from, if in the first second the cargo fell 4.5 meters, and in the following seconds it fell 9.8 meters more than the previous second?

336) Prove an identity.

$$(n-3) \times (n-2) + 19 - 5n = n^2 - 20n + 25$$

337) Ginny took some time off from study. She slept for $1\frac{1}{15}$ hours, read a book for $3\frac{1}{6}$ hours, and spent $2\frac{1}{4}$ hours at the movie. How long did Ginny rest?

338) Multiply.

$$(4x - 3y + 2z) \times (3x - 3y)$$

339) Steve ran a 3-day marathon. The first day he ran 2/5 of the distance, the second day – 1/3 of the distance. What part of the planned marathon route does Steve have to run on the third day?

340) Write an expression in factored form.

$$ax^2 - ay - bx^2 + cy + by - cx^2$$

341) Robert invested money into a risky stock and lost 28%. How much money did he invest, if he had $288 left?

342) Simplify the expression.

$$(6a^3 - 4ab^2 + 1) \times \frac{1}{2}ab$$

343) How many pounds of cherries have to be used to make 16.4 pounds of sugar-free jam, if cherries loose 18% of their weight in processing?

344) Perform the indicated operations.

$$\frac{1}{14x^2} - \frac{1}{21xy} - \frac{1}{7y^2}$$

345) A furniture store has 25 desks for sale. Some of these desks are 3-drawer desks, and others are 4-drawer. If there are total of 91 drawers in all desks, how many desks of each type are in the store?

346) Divide and express the quotient in lowest term.

a) $-6.4xy \div (-4x)$

b) $(-0.24abc) \div (-0.6ab)$

347) The New Horizons construction company built 72 two-bedroom and three-bedroom condominiums. How many of each type were built, if the total number of bedrooms is 168?

348) Perform the indicated operations.

$$\frac{a-9ab}{(b-2)(b+2)} \times \frac{b+2}{1-9b}$$

349) A wheel makes $27\frac{5}{6}$ complete rotations in one minute. How many rotations will the wheel make in 3 minutes? In $1\frac{1}{4}$ minutes? In $\frac{2}{3}$ minutes?

350) Simplify.

$$\left(\frac{1}{x-y} - \frac{1}{x+y}\right) \times \frac{x+y}{2y}$$

351) There are $27\frac{1}{5}$ pounds of copper in the copper-tin alloy. If tin is $\frac{3}{17}$ of the weight of the copper, find the total weight of the alloy.

352) Simplify the expression.

$$(a - 2b)(a - 2b) - (a^2 + 4b^2)$$

353) A birch tree lives for 150 years. A pine tree lives $2\frac{1}{3}$ times longer. An oak tree lives 5 times longer than a pine tree. How long does the oak live?

354) Perform the indicated operations.

$$\frac{7b^4}{9c^5y} \div \frac{35b^4c}{18c^4y^2}$$

355) The sum of the numerator and denominator of a fraction is 23. The numerator is 9 less than the denominator. Find the fraction.

356) Multiply.

$$(3a - 3b + 4c) \times (3a - 5b)$$

357) The sum of the numerator and denominator of an improper fraction is 27 and the difference is 13. Find the fraction.

358) Perform the indicated operations.

$$\frac{a^2 + ab}{(a-b)(a+b)} \div \frac{a+b}{ab-b^2}$$

359) A man wishes to divide his herd of 21 horses between his three sons so that the oldest son gets $1\frac{5}{7}$ times more than the middle son, while the youngest son gets $\frac{2}{7}$ of the number of horses the middle son gets. How many horses does each son get?

360) Simplify.

$$(n-1)n(n+1)+1$$

361) Tussah won $5,600. She wants to spend part for a trip, part to make repairs on her house, and part for her savings account. The amount set aside for the trip is 1/7 of the amount for the house repair and the amount for the house repair is 7/8 of the amount for a savings deposit. How much money does she put in savings, for house repair, for the trip?

362) Find the numerical value of the expression.

$$(a-1)(a-2)+(a-3)(a-2). \text{ Use } a = 0.2$$

363) Dora has two piggybanks. The first one contains 1.5 times less coins than the second one. If Dora takes 5 coins from the first piggybank and put them in the second one she will have 2 times more coins in the second piggybank than in the first one. How many coins were in the each piggybank originally?

364) Divide and express the quotient in lowest term.

$$\frac{1}{3}m^3n^2p^2 \div (-\frac{2}{3}m^2n^2p^2)$$

365) There were 2.5 times more people on a cross-town bus than on a downtown bus. When 5 people transferred from the downtown bus to the cross-town bus, the downtown had 3 times fewer passengers than the cross-town bus. How many people were on the each bus at the beginning?

366) Simplify.

$$\frac{1}{1-b} + \frac{1}{1+b} - \frac{2+b}{(1-b)(1+b)}$$

367) 6-th graders organized a Christmas Food Drive. They planned to collect 2,500 cans of food, but instead they had 3,200. What percent of cans were collected compared to that planned? By what percentage did 6-th graders exceed the planned number of cans?

368) Perform the indicated operations.

$$\frac{c+d}{6(2c-d)} \div \frac{(c-d)(c+d)}{2c-d}$$

369) A survey shows that 5/7 of all households in the Creek town have dogs, the rest have cats. How many times more households have dogs than cats?

370) Multiply.

$$n(n+1)(n+2)\times 5$$

371) Linda is a waitress at the Blue Water restaurant. On Saturday she made $60 in tips, which amounted to 2/29 of all the money paid by her customers. How much money did customers pay at Linda's tables?

372) Simplify an expression.

$$0.6ab^2 + [2a^3 + b^3 - (3ab^2 - (a^3 + 2.4ab^2 - b^3))]$$

373) Each weekend Chris earns money delivering groceries. He deposits 5/24 of what he earns in his savings account; keeps $24 for spending money and the rest goes to pay off his car. How much does Chris earn if the amount of money he keeps is 4/5 of the amount he puts into savings?

374) Factor each expression completely.

a) $(b+1)^2 - b - 1$

b) $(a-1)^2 + 2(a-1)$

375) The average of three numbers is 8.9. The second number is greater than the first one by 0.7. The third number is greater than the second by 0.7. Find all the numbers.

376) Perform the indicated operations.

$$\frac{16x^2 y}{7z} \div \frac{10xy^3}{21z^2}$$

377) Danny received $5.10 in dividends on 21 shares of stock. How much should Andrew receive in dividends on 7 shares of the same stock?

378) Factor the expression.

$$p^2 - 2p + p(p-2)^2$$

379) How much silver is in 150 kilograms of an alloy that has 28% of metals other than silver?

380) Find the solution set for the equation.

$$\frac{4x-3}{2} - \frac{5-2x}{3} - \frac{3x-7}{6} = 0$$

381) In a factory, 200 parts were made. When they were tested 170 were found to be acceptable. What was a percentage of acceptable parts made?

382) Find the numeric value of the expression. Use $x = \frac{5}{6}, y = \frac{2}{3}$.

$$\frac{5x-5y}{(x+y)(x-y)} \times \frac{3x+3y}{10y-10x}$$

383) Laura answered 95% of the questions on a test correctly. If she had 57 correct answers, what was the total number of questions on the test?

384) Find the solution set for each equation.

a) $7 - \dfrac{x}{2} = 3 + \dfrac{7x}{2}$

b) $9 - \dfrac{2x}{3} = 7 + \dfrac{x}{3}$

385) The scale on a map is given as 3.6 centimeters to 72 kilometers. How far apart are two towns if the distance between these two towns on the map is 12.6 centimeters?

386) Factor the expression.

$$4c^4d + 8c^2d^2 + 12c^2d$$

387) The distance between two gas stations is 3 kilometers which, on a map, is 6 centimeters. How far apart are two towns on the same map if the distance between them is 10 kilometers?

388) What value of x makes expressions $\frac{1}{2}(x-7)+1$ and $\frac{3(1-x)}{4}$ equal?

389) There are about 60 calories in 100 grams of tuna. About how many calories are there in a 650-gram slice of tuna?

390) Factor each expression completely.

   a) $8a^2b^2 + 24a^3b^3$
   b) $(4a-1)^2 + 2(4a-1)$

391) The tank of one car has 104 liters of gas, the tank of a second car has 72 liters. If the first car uses 3 liters per hour of gasoline and the second one–5 liters per hour, how soon would the first tank have 2.5 times more gas remaining than the second car?

392) Perform the indicated operations.

$$\frac{4}{6b-36}+\frac{1}{6-b}$$

393) A cyclist travels $7\frac{1}{2}$ kilometers in $\frac{3}{5}$ hours. How many kilometers will he go in $2\frac{1}{2}$ hours traveling at the same rate?

394) Factor the expression.

$$25a^2 - 10ab + 30ab - 12b^2$$

395) It takes Amy 20 kilograms of apples to make 16 kilograms of apple jam. How much jam could she make if she had 45 kilograms of apples?

396) Find the solution set for the equation.

$$\frac{2x-3}{2} - \frac{3-4x}{4} - \frac{3-5x}{8} = 0$$

397) You earn $4.75 for every $250 of magazine subscription you sell. How much money do you earn selling $550.00 subscription?

398) Factor each expression completely.

a) $4ab^2 + 15abc - 4bcd - 15c^2d$

b) $m^3 - m^2 + m - 1$

399) There is 18% of sand in a certain concrete mixture. How many tons of sand are in 33.5 tons of the same mixture?

400) What value of x makes expressions $\dfrac{2}{5}(3-2x)$ and $\dfrac{3(1-x)}{10}-\dfrac{4}{5}$ equal?

401) There is 45% of oil in a certain type of sunflower seed. How many kilograms of oil are in 29.2 kilograms of seeds?

402) Factor the expression.

$$25a^3 + 50a^2b - 9b^2a - 18b^3$$

403) A recipe calls for 0.4 kilograms of sugar for 6 kilograms of flour. What is the percentage of sugar in the mixture?

404) Perform the indicated operations.

$$\frac{46d^3c}{15a} \div \frac{23dc^3}{5a^3}$$

405) Ann owns a candy store. There are $8\frac{1}{10}$ pounds of chocolate candies in the first box, $2\frac{2}{9}$ times more pounds of lollypops in the second box, and $1\frac{1}{8}$ times less pounds of mints than lollypops in the third box. How many pounds of lollypops and how many pounds of mints does Ann have in the store?

406) Perform the indicated operations.

$$\frac{4}{16-a^2} + \frac{2}{4+a}$$

407) On Tuesday, August 13, 1996, the price of a hamburger chain's stock closed at $4\frac{9}{10}$. The previous Wednesday, the stock closed $1\frac{2}{5}$ times lower than on Tuesday and on previous Friday, the stock closed $2\frac{2}{7}$ times higher than on Wednesday. At what price did the stock close on Friday?

408) Find the solution set for the equation.

$$\frac{x+4}{5} - \frac{x+3}{3} = \frac{x-5}{5} + \frac{x-2}{3}$$

409) For the first 3 days of a week, the snowfall in Chicago was 28.2 inches. On the first day of the week, 0.3 inches less snow fell than on the second day. On the third day of the week, 0.9 inches more snow fell than on the second day. How many inches of snow fell each day?

410) Perform the indicated operations.

$$\frac{18m^3n^5}{7k} \div (9n^2)$$

411) 1.6 pounds of lamb contain 0.4 pounds of protein. How many pounds of protein are in 3.2 pounds of lamb?

412) Simplify.

$$\frac{1}{x} + \frac{3}{5xy} - \frac{5}{4y}$$

413) 6.5 kilograms of pork contain 2.6 kilograms of fat. How many kilograms of fat are there in 10.5 kilograms of pork?

414) Factor the expression.

$$2a + b - cb^2 - 2acb$$

415) A motorboat travels 96 miles upstream and 240 miles downstream. If the river flows at the rate of 4 miles per hour and the motorboat moves at 20 miles per hour, what is the average speed of the boat?

416) Find the numerical value of the expression.

$$(a-4)(a-2)-(a-1)(a-2). \text{ Use } a=1\frac{3}{4}$$

417) A ferry sails for 240 miles from one port to another, going along with the current and then travels back. If the rate of the current is 2 miles per hour and the rate of the ferry is 18 miles per hour, what is the average speed of the ferry?

418) Divide and express the quotient in lowest term.

$$(-1\frac{1}{2}a^4b^3c^2) \div (-\frac{2}{3}a^2b^2c^2)$$

419) One pipe can empty $\frac{3}{8}$ of a tank in 9 hours. How much time does this pipe require to empty $\frac{7}{12}$ of the same tank?

420) Solve the equation.

$$\frac{2x-4}{5} + \frac{2x-1}{3} = 1$$

421) Two painters could paint a house in $2\frac{2}{5}$ days. If the first painter alone could do it in 4 days, how long would it take the second painter?

422) Find the solution set for the equation.

$$\frac{2x}{3} - \frac{1-3x}{5} = \frac{x-7}{15}$$

423) The freshman class earned $700.80 for their annual trip to the amusement park by washing cars, selling cookies at a bake sale, and giving concerts. The amount of money they made selling cookies was 50% of what they made washing cars. The band earned 1.5 times more money than the car wash. How much money did the class make for each activity?

424) Simplify.

$$\frac{a-1}{a} + \frac{a-1}{2a} - \frac{3}{2}$$

425) Jaime paid \$113 for three sweaters with his high school logo. Sweaters were made of cotton, wool, and acrylic. The price of a cotton sweater was $\frac{4}{5}$ the wool sweater and a wool sweater was 70% of a price of an acrylic sweater. How much did Jaime pay for each sweater?

426) Solve for the variable in each equation and check.

a) $2(x-1) = 3(2x-1)$

b) $3(1-x) = 4x - 1$

427) A boat travels 32.1 miles going downstream in $\frac{3}{4}$ hours. The same boat goes 30.56 miles against the stream in 0.8 hours. Find the boat's own speed.

428) Factor each expression completely.

$$\text{a) } (a+3)-6(a+3)+9(a+3)$$
$$\text{b) } (m-1)(m-7m)+(m-1)(5m+1)$$

429) A piece of silver-copper alloy weighs 2 pounds. If the content of silver in the alloy is $14\frac{2}{7}$ % of that of the copper, what is the weight of the copper in the alloy?

430) Factor completely.

$$a^3 - 8a^2 - 3a^2 + 24a$$

431) The Petal Department Store chain was closing one of its stores so they had to sell most of the inventory in the store. In the final sale, they marked all items $33\frac{1}{3}$ % off the ticket price. What was the original price of a sweater if it sells now for $46.20?

432) Solve the equation.

$$5 - \frac{2x-5}{3} = \frac{4x+2}{3}$$

433) A jar contains 88 red and green jellybeans. The number of red jellybeans is $2\frac{2}{3}$ times less than the green ones. How many of each color jellybeans are in the jar?

434) Perform the indicated operations.

$$\frac{3}{a} + \frac{2}{b} + \frac{a+b}{ab}$$

435) There are 26 construction workers in two teams. How many people are in each team, if the number of workers in the first team is $1\frac{1}{6}$ times more than that in the second team?

436) Perform the indicated operations and simplify.

$$\left(m - \frac{1}{m}\right) \div \left(m + \frac{1}{m}\right)$$

437) A 12-pound piece of copper-tin alloy contains 10.8 pounds of copper. How much copper would 10 pounds of the same alloy contain?

438) Solve the equation.

$$\frac{3x + 7}{2} + \frac{3 + x}{3} = 5$$

439) Two tanks contain 725 liters of gasoline. When 1/3 of gasoline was used from the first tank and 2/7 from the second tank, both had the same amount of gasoline. How much gasoline was in each tank originally?

440) Multiply.

$$49a^2bc^2 \bullet \left(-\frac{2}{7}ab\right) \bullet \frac{1}{14}ac$$

441) The sum of two numbers is 75. 2/3 of the first number is equal to 4/9 of the second. Find both numbers.

442) Solve for the variable in each equation and check.

$$\text{a) } 3-5(x-1)=x-2$$
$$\text{b) } 3(x-2)-2(x-1)=17$$

443) Do you believe in love at first sight? In a survey, 3/8 of those asked said yes, $\frac{1}{4}$ said, maybe, and 21 said, no. How many people were asked?

444) Simplify the expression.

$$\left(\frac{1}{m}\right)^3 \bullet \left(\frac{1}{m}\right)^2 \bullet m^5$$

445) How many marbles are in the bag, if 60 are green, 4/9 of all marbles are red, and 1/3 are blue?

446) Factor completely.

$$9b^2 - 27b - b + 3$$

447) Andy spent 3/7 of all his money on a book and 5/14 on a notebook. How much money did he have originally if the book cost $7 more than the notebook?

448) Multiply in each case.

a) $(a+b) \times (7a - 5b)$

b) $(4ab - 1) \times (3a - b)$

449) A woman went to a department store and bought a dress among other things, paying for the dress 3/5 of what she spent in the store. The next day she went to the store again and bought the same dress for the same price for her sister, paying 9/10 of money spent on that day. How much did the dress cost if she spent a total of $750 in the two days?

450) Solve the equation.

$$\frac{x-5}{5} = \frac{2x+1}{3} - 7$$

451) Is it true that:

a) a sum of two prime numbers is a prime number?

b) a product of two prime numbers is a prime number?

c) a product of two non-prime numbers is a non-prime number?

452) Simplify the expression.

$$\frac{(b^3)^2 b^3 b}{(b^2)^4} - b^2$$

453) The sum of two numbers is 177. If you divide the bigger of the two by the smaller the quotient is equal to 3 and the remainder is 9. Find both numbers.

454) Solve each equation.

a) $\dfrac{2x+1}{3} = 6$

b) $\dfrac{x-7}{2} = \dfrac{1}{4}$

455) Mr. Chants owns a car dealership with two different car makes. There are 60 cars on the lot. He sells the first make for $13,000 and the second make for $15,000. How many cars did he have originally, if he made $220,000 more in sales on the first make?

456) Simplify the expression.

$$(1-a)(1+a+a^2)+a^3$$

457) Before Christmas Andy's and Brown's department stores received a stock of late fashion coats. Andy's got 110 coats and Brown's got 130. After Brown's sold twice more coats than Andy's, Andy's still had 5 coats more in stock than Brown's. How many coats did each store sell before Christmas?

458) Simplify the expression.

$$\left(\left(\frac{1}{a}\right)^4\right)^3-\left(\frac{1}{a}\right)^{11}\bullet\frac{1}{a}$$

459) Anthony was rafting downstream for 3 hours at 4 miles per hour when his friend Joe started to follow Anthony on a motorboat. How soon did Joe overtake Anthony if his boat was going at 9 miles per hour?

460) Simplify the expression.

$$\frac{(3b^2)^2 3^2 b^3}{3^4 b^6} + b$$

461) A train covers a distance of 330 miles in $2\frac{3}{4}$ hours. What distance would the same train cover traveling at the same rate for 7.5 hours?

462) Simplify the expression.

$$(b+3)(b+9) - 27$$

463) Cheryl has $25.50 in her wallet. If she spends 3/5 of her money on 5 feet of fabric, how much would one foot cost?

464) Multiply.

$$-12a^2bd \times 5ad \times (-b^2d^2)$$

465) Working together, Mary and Fanny can sew an evening dress in 8 days. If it takes Fanny 12 days to do the job alone, how long would it take Mary to make a dress working by herself?

466) Solve each equation.

a) $\dfrac{x}{3} - \dfrac{1}{2} = \dfrac{x}{2}$

b) $\dfrac{4}{3}x - 1 = \dfrac{x}{9} + \dfrac{1}{6}$

467) Two snowplows working together can clean up the snow in New City in 6 hours. After 3 hours of work the first plow was sent to another region to help and the second plow finished the work in 5 hours. How long would it take for each snowplow to do the work alone?

468) Multiply in each case.

   a) $(a + 3b - 4c) \times (a - 3b - 4c)$
   b) $(m + n - 2) \times (m - n + 2)$

469) Ann can type a manuscript in $3\frac{1}{3}$ days. Rita can type the same manuscript in $2\frac{1}{2}$ days. How long would it take for Ann and Rita to finish typing the same manuscript working together?

470) Simplify the expression.

$$\left(2 - c^2\right) \times \left(\frac{1}{4} + \frac{1}{2}c\right)$$

471) Green College School of Agriculture' students are growing corn, wheat and sunflowers. The total area of the experimental field is 52.5 square miles. The cornfield occupies 2/5 of the total area, while the wheat field takes up 1/3 of the total area, and the sunflowers field occupies the rest. How many square miles does the sunflower field take up?

472) Multiply in each case.

a) $\left(\dfrac{1}{3} a^2 b - \dfrac{2}{5} ab^2 \right) \bullet (15a - 30b)$

b) $\left(\dfrac{1}{2} a^2 + 4a + 1 \right) \bullet (3a - 1)$

473) Mr. Kraus spent \$720.80 on clothes. He paid $\dfrac{1}{4}$ of that for a suit and spent the rest to buy 12 shirts of equal price. How much did each shirt cost?

474) Perform the indicated operations.

$$\dfrac{x+y}{x} \div \left(1 + \dfrac{y}{x} \right)$$

475) During road construction in Oregon, $\dfrac{8}{27}$ of it was built during the first month, $\dfrac{4}{9}$ was built in the second month, and the last $5\dfrac{1}{4}$ miles were built during the third month. What is the total length of the road?

476) Factor completely.

$$16b^2a - 4b - 8ab^2 + 2b$$

477) Mary spent 2/7 of all her money at Marcy's Department store, and 3/5 of the rest at the Right's. How much money did she have originally, if she still had $60 left?

478) Find the solution set of each equation.

a) $\dfrac{x+3}{2} = x - 4$

b) $2 - 3x = \dfrac{x-12}{2}$

479) Two typists had to type a manuscript. The first typist did 3/7 of the assignment and the second typist did 5/14. If the first one typed 7 pages more than the second typist, how many pages were in the manuscript?

480) Simplify the expression.

$$(a+1) \bullet (a-1) \bullet (a^2 +1)$$

481) Peter, Andrew, and Sam have earned money to buy equipment for their band. They have already spent $\frac{3}{10}$ of their money on a used guitar and drum set, $\frac{1}{2}$ on an amplifier, and the last $600 on promotional materials and T-shorts for the band. How much money did they originally earn?

482) Simplify.

$$\frac{m-nm}{4m} - \frac{mn-n}{5n}$$

483) Carl laid 3/8 of the bricks that were delivered to the construction site; Tony laid 2/5 of all of the bricks. How many bricks were delivered to the construction site if Tony laid 840 bricks more than Carl did?

484) Factor completely.

$$-a^2 - 2a - b^2a^2 - 2b^2a$$

485) A truck leaves town A going to town B at a constant rate of 60 miles per hour. 2 hours later a car leaves town B going towards the truck at a speed $1\frac{1}{2}$ times faster than the truck. How long will it take them to meet if the distance between towns is 450 miles?

486) Find the numeric value of the expression. Use $a = \frac{3}{4}, b = \frac{1}{9}$.

$$12a^2b^3 \div (3ab^2)$$

487) A freight train and a passenger train start towards each other at the same time from two towns that are 169 miles apart. The freight train covered 5/8 of the distance the passenger train covered before two trains met. What distance did each train cover before they met?

488) Simplify the expression.

$$(2ab + 3) \times (3 - 2ab) + 4a^2b^4$$

489) John rides his bicycle from Syracuse traveling west. At the same time, Martin leaves a village 62 kilometers west of Syracuse, walking east at a constant rate. Martin covers 11/20 of the distance John covered before they meet. How long would it take John to meet Martin if the rate of the bicycle is 4.5 kilometers per hour more than the rate of the pedestrian?

490) Perform the indicated operations.

$$24k^2 \div \frac{12m^4k^2}{11p^3n}$$

491) There are 32,000 cubic yards of earth that must be removed from the excavation site. 60% of earth has to be loaded on trucks and removed from the site. How many cubic yards of earth stayed on the site?

492) Perform the indicated operations.

$$\frac{10a + 6b}{2(a - b)} + \frac{3a - 2b}{a - b}$$

493) Mr. Karp spent 20% of his bonus on a new $440 bicycle. What is Mr. Karp's salary, if the size of his bonus is 12.5% of it?

494) Find the numeric value of the expression. Use $m = \frac{1}{7}, n = 1.$

$$(-49m^3n^4) \div (7mn^4)$$

495) After the upgrade of the milling machine its productivity was up 10%. After the second upgrade it was up another 10%. What was the total increase in productivity after two upgrades?

496) Factor completely.

$$-9 + 6b^2 - b + 54b$$

497) The print shop used 60% of total paper stock in two days. The amount of paper used on the second day was $1\frac{1}{5}$ times more than on the first day. How much paper was used on the first day, if the total stock of paper was $6\frac{3}{5}$ tons?

498) Simplify the expression.

$$(1 - 2b) \times (1 + 2b) \times 15$$

499) Mr. Rose invested 75% of his money in two business enterprises. In one enterprise he invested 5/7 of the amount invested into the second enterprise. How much did Mr. Rose invest into each enterprise if he had $3,040 originally?

500) Simplify the expression.

$$(a+3)^2 + (a-3)^2$$

501) One pipe fills an entire tank in 3 hours. If it fills 5/18 of the total volume in the first hour, 1/3 in the second hour, and 10 gallons more in the third hour than in the second hour, what is the tank's total volume?

502) Find the numeric value of the expression. Use $a = -1, b = 5.$

$$(4a^3b + 6a^2b) \div (2a^2b)$$

503) The difference of two numbers is 72. Find both numbers if 4.5% of the first number is equal to 8.5% of the second one.

504) Simplify the expression.

$$\frac{1}{3}\times(4a^4-\frac{2}{3}a^2+\frac{1}{9})-\frac{1}{27}$$

505) A dealer bought three birds for $85.50. The first bird cost 75% of the money paid for the second bird and the third bird cost 110% of the amount paid for the second bird. How much did each bird cost?

506) Perform the indicated operations.

$$\frac{7-a}{a+b}\div\frac{a-b}{a+b}$$

507) A salesman traveled 60 miles to see a customer. During the first hour of the trip he covered a distance, 2/3 of which was 3.5 times less than the number of miles left to cover. How many more miles did the salesman have to travel?

508) Perform the indicated operations.

$$\frac{a+4b}{2b} - \frac{4b}{2b-a}$$

509) Farm workers set out to plow a 240-acre field. After working for two days they plowed part of the field, 80% of which was 2.5 times less than the amount of acres left to plow. How many days did it take to plow the whole field?

510) Multiply.

$$\left(-\frac{1}{2}a^4b^2c\right) \times \frac{2}{15}abc^3$$

511) There are a total of 360 planes at two airbases. All of the planes at the first base, a number equal to 80% of the planes at the second airbase, took off on a mission. How many planes were at each airbase?

512) Simplify the expression.

$$4a^2 - (2a - b)(2a + b)$$

513) Mike and Nick have $28. When Mike spends 75% of his money and Nick spends 2/3 of his money, they both have the same amount of money left. How much money did each boy have originally?

514) Find the numeric value of the expression. Use $a = \dfrac{1}{4}$

$$(12a^4 - 24a^3 + 12a^2) \div (6a^2)$$

515) A car finished its journey in 1 hour and 15 minutes. If the car traveled 10 miles per hour slower, it would has finished the trip in 1 hour and 30 minutes. Find the rate of the car and total distance covered.

516) Perform the indicated operations.

$$\frac{x-y}{2a} \times \frac{4b}{x-y}$$

517) There are 840 liters of water in the first tank and 4/7 of that in the second tank. The first tank is being emptied at the rate 3 times faster than the second one. In 5 minutes the first tank would have 40 liters less than the second tank. How much water pours out of each tank in one minute?

518) Find the numeric value of the expression. Use $a = 1$

$$\frac{a+5}{a-3} \div \frac{a+5}{9-a^2}$$

519) Alex and Carry start toward each other at the same time from two villages that are 21 kilometers apart. Alex walks at the speed rate of 6 kilometers per hour and covers $1\frac{1}{3}$ times more distance than Carry does before they meet. How much time have they spent on the road before they meet?

520) Perform the indicated operations.

$$\frac{a^2 + b^2}{a^2 - b^2} - \frac{a^2 - b^2}{2(a^2 - b^2)}$$

521) A motorboat covers a certain distance going against the stream in 15 hours. Traveling downstream the same motorboat covers the same distance in 13 hours. Find the rate of the current if the speed of the boat is 70 miles per hour?

522) Simplify the expression.

$$(a - 5) \times (5 + a) + 25$$

523) The average (arithmetic mean) of 4 integers is equal to 25.5. Beginning with the second number each one of them is greater than the previous one by 7. Find all numbers.

524) Perform the indicated operations.

$$\frac{c+d}{c-d} \div \frac{c\times(c+d)^2}{c-d}$$

525) Leo drove a car for 3 hours at 53.5 miles per hour, 2 hours at 62.3 miles per hour, and 4 hours at 48.2 miles per hour. What was his average driving speed during the trip?

526) Prove the identity.

$$(a-b)^2 = (b-a)^2$$

527) A Titanium Movie Theater was full for the movie premiere. How many tickets were bought if 1% of the number of all tickets sold is equal to 7?

528) Simplify the expression.

$$(2a+b)^2 + (2a-b)^2$$

529) Ann started from town A cycling at 10 kilometers per hour. Her friend Tamara started from the same town two hours earlier walking at 5 kilometers per hour. How long would it take for Ann to overtake Tamara?

530) Add.

$$a^2 + ab + b^2 + (2a^2 + 3ab - 2b^2) + (a^2 + ab + 2b^2)$$

531) Two trains traveling toward each other at the steady rate of 80 miles per hour and 90 miles per hour, respectively, were 850 miles apart at the start. How long will it take for the trains to meet?

532) Perform the indicated operations.

$$\frac{a-b}{2b} \div \frac{a-b}{8b^3}$$

533) A train started from Chicago going west at 48 miles per hour. Two hours later another train started from the same terminal going east. In 3 additional hours the distance between the trains was 402 miles. Find the rate of the second train.

534) Factor the expression completely.

$$a^2 - 2a + 6a - 3a^2$$

535) Two trains left the station at the same time traveling in opposite directions. The first train traveled 18 miles per hour faster than the other. How long would it take for both trains to be 504 miles apart, if the rate of the second train was 54 miles per hour?

536) Find the numeric value of the expression.

$$\frac{3n - 3m}{n + p} \div \frac{6m - 6n}{n + p}$$

537) Two bicyclists started towards each other from a distance of 76 miles. They met in 2 hours. Find the speed rate of each bicycle, if the first one traveled 3 miles per hour slower than the second one.

538) Prove the identity.

$$(-a - b)^2 = (a + b)^2$$

539) Five small and two big boxes contain 54 crayons, at the same time 3 small boxes and 2 big boxes contain 42 crayons. How many crayons are in a big box? How many crayons are in the small box?

540) Perform the indicated operations.

$$\frac{a^2 - ab}{b} \times \frac{b^2}{a}$$

541) A Lexus sedan without its passengers is 1,125 tons heavier than its passengers. The passengers weigh 6 times less than the car. Find the weight of the car and the weight of the passengers separately. (1 ton = 1000 kilograms)

542) Find the numeric value of the expression. Use $a = 2.5$.

$$\frac{a^2 - b^2}{3a + 3b} \times \frac{3a^2}{5b - 5a}$$

543) One number is 3 times larger than the other number. If 46 is added to the first number and 18 is added to the second number, the sum of both numbers is 184. Find the numbers.

544) Multiply.

$$\left(-\frac{4}{3}m^5n^3\right)\times\left(-\frac{3}{4}mn^3\right)$$

545) A grandfather is 6 times as old as his grandson. The grandson is 3 times as young as his father. The grandfather is 55 years older than the grandson. Find the age of all three men.

546) Simplify.

$$\left(\frac{2}{2-p}+\frac{1}{p+2}\right)\div\frac{p+6}{p+2}$$

547) There are 75 pages in the book. The first day Andrew read 3/5 of the book and the next day he read 2/5 of the remaining pages. How many pages does he have left to read in order to finish the book?

548) Factor the expression completely.

$$b^2 - 7b + 14b - 2b^2$$

549) There are 500 students in the town high school and 2/5 of that number in the town middle school, while in the town elementary school there are 8/5 as many students as there are in the middle school. How many students are in the elementary school?

550) Prove the identity.

$$(-a - b)(a + b) = -(a + b)^2$$

551) A milk shake recipe calls for 560 grams of milk, which is 7/10 of the total weight of the shake; 3/20 of the total weight is sugar, 1/10 is vanilla extract, 1/20 is cocoa. How many grams of each ingredient are used?

552) Solve each equation and check.

a) $10 - 0.3b = 6.4$

b) $b \div 10 - 0.1 = 0.07$

553) To write an essay, Gene used up 5/8 of his notebook. How many pages are in the notebook if there are still 36 pages left unused?

554) Perform the indicated operations.

$$\frac{ab + b^2}{9} \div \frac{b^2}{3a}$$

555) Kathy would pay \$12 more for $\frac{1}{2}$ of a yard of fabric than if she purchased $\frac{1}{3}$ of a yard of the same fabric. Find the price of a yard of fabric.

556) Simplify the expression.

$$\frac{5(a-b)}{3(a^2+b^2)} \div \frac{(a-b)^2}{a^2+b^2}$$

557) Mary's father asked her to paint $\frac{2}{5}$ of a fence. Mary asked her sister Kerry to help out. Kerry completed $\frac{1}{4}$ of what Mary had to paint. What is the total length of the fence, if Kerry painted $2\frac{1}{2}$ meters? How many meters of the fence did Mary paint?

558) Add.

$$2a^2 + 2ab + 3b^2 - a^2 - 2b^2 - 4ab + 6a^2 - 10b^2$$

559) Agnes bought a box of pastries at the Italian bakery. She gave half of them to her friends on the bus. At home, she gave $\frac{1}{4}$ of the remaining pastries to her sister, 1/4 to her mother, and had two pastries left. How many pastries did Agnes buy?

560) Transform each expression into a polynomial.

        a) $(x+1)^2$

        b) $(5z-t)^2$

561) Anna had some math problems to do for homework. She did 1/2 in the study room at school and 2/3 of those remaining at the school library. She had 3 to finish at home. How many math problems was she assigned for homework?

562) Simplify.

$$\frac{2a}{2a+b} \div \frac{4a^4}{-b-2a}$$

563) To make pillowcases, Joan cuts several 0.65-meter-long pieces of fabric from a 5.5-meter-long roll. How many pillowcases did Joan make, if there is a 0.95-meter-long piece of fabric left?

564) Simplify the expression.

$$(4a + b)^2 - (4a - b)^2$$

565) One bag contains 37.5 pounds of sand that is 1.5 times heavier than the second bag and 12.5 pounds lighter than the third bag. How many pounds of sand are in each bag?

566) Transform each expression into a polynomial.

a) $(c + d)^2$

b) $(q + 2p)^2$

567) There are 1.5 pounds of coffee in one container and 0.9 pounds of coffee in the other. How much coffee should be moved from the first container into the second one in order to have the same amount of coffee in each container? How much coffee will then be in each container?

568) Prove the identity.

$$(a-b)^2 = (b-a)^2$$

569) To manufacture 15 pairs of pants one needs 18 meters of fabric. How many meters of fabric will be left if only 8 pairs are made?

570) Simplify first and find the numeric value of the expression. Use $m = 142, n = 42$.

$$5m^2 - 10mn + 5n^2$$

571) Station A is located 165 miles from station B. A train travels from A to B in 1.5 hours at a rate of 60 miles per hour. What should the rate of the train be to cover the remaining stretch in 1.2 hours?

572) Simplify the expression.

$$\frac{1-a}{3b^2} \bullet \frac{b^3}{1-a^2}$$

573) Linda earns a monthly salary of $2,325, plus a commission of 8% of her total sales for the month. Last month her total earnings were $3,500. What were her total sales?

574) Simplify.

$$\frac{2q}{p+2q} \div \frac{2q^2}{(p-2q)(p+2q)}$$

575) Lenny paid $10,584 for a used car, which included an 8% sales tax on the base cost of the car. What was the cost of the car, without the sales tax?

576) Transform each expression into a polynomial.

       a) $(x - y)^2$

       b) $(3x + 2y)^2$

577) Ronny goes for a walk at a speed of 3 miles per hour. Two hours later Martin attempts to over-take him by running at the rate of 7 miles per hour. How long would it take him to reach Ronny?

578) Simplify first and find the numeric value of the expression. Use $m = 56, n = 44$.

       $6m^2 + 12mn + 6n^2$

579) Two cars leave town at the same time and travel in opposite directions. One car travels at the rate of 50 miles per hour, and the other at 55 miles per hour. In how many hours will the two cars be 420 miles apart?

580) Simplify the expression.

$$\frac{5m}{m^2 - n^2} \div \frac{15m^2}{m - n}$$

581) Dahlia has $1.69 in pennies, nickels, and dimes. She has twice as many pennies as she has nickels, and five more dimes than nickels. How many coins of each type does Dahlia have?

582) Simplify the expression.

$$\left(\frac{2}{a} - \frac{a}{b}\right) + \frac{a}{b}$$

583) Kim walks at 2.7 miles per hour for 0.6 hours. How much time would it take her to ride her bicycle the same distance, if she can ride 2.4 times faster than she can walk?

584) Simplify the expression.

$$\frac{3(x-y)}{4y^2(x^2+y^2)} \times \frac{x^2+y^2}{x-y}$$

585) The sum of four numbers is 6.2. The first number is equal to the second one, the third one is 3 times as large as the fourth one, and the fourth one is 0.8. Find each number.

586) Simplify.

$$\left(\frac{1}{1-a}-1\right) \div \left(1+\frac{1}{1-a}\right)$$

587) To make a coat for a 4-year old girl one needs 1.4 meters of fabric. It takes 0.2 meters less to make a skirt for her. A garment factory received enough wool to make some coats and skirts. They used 1/3 of all the fabric delivered to make 60 coats and the rest to make skirts. How many skirts can be made?

588) Simplify the expression.
$$(1-7b)^2 - (1+7b)^2$$

589) The combined weight of two wedges of cheese is 1.4 kilograms. Find the weight of each wedge, if one wedge is 3 times as heavy as the other one.

590) Transform each expression into a polynomial.

a) $(2+x)^2$

b) $(6a-4b)^2$

591) There are 2.75 pounds of apples in two shopping bags. How much does each bag of apples weigh if the first one is 1.5 times lighter than the second one?

592) Simplify first and find the numeric value of the expression. Use $a = 2, b = 9.$

$$-36a^3 + 4a^2b - \frac{1}{9}ab^2$$

593) What number must be subtracted from both the numerator and denominator of the fraction 11/15 to get a fraction whose value is 3/5?

594) Factor each polynomial completely.

a) $-4 + (a - 2)^2$

b) $(1 - a)^2 - 1$

595) The denominator of a fraction is 2 more than the numerator. If 4 is added to the numerator and 3 is subtracted from the denominator, the value of the new fraction is 6. Find the original fraction.

596) Transform the expression into a polynomial.

$$(0.2x + 0.3y)^2$$

597) One pipe can empty a tank in 3 hours. A second pipe takes 4 hours to complete the same job. How long will it take to empty the tank if both pipes are used?

598) Simplify first and find the numeric value of the expression. Use $a = -6, b = 4$.

$$-4a^3b - 8a^2b - \frac{1}{4}ab^2$$

599) A student received grades of 72, 75, and 78 on three tests. What must her score on the next test be for her to have an average grade of 80 for all four tests?

600) Simplify first and write the number that the numerical expression represents.

$$101^2 - 2 \times 101 \times 81 + 81^2$$

601) Andy takes twice as long as Tom to complete a certain job. Working together, they can complete the job in 6 hours. How long will it take for Andy to complete the job by himself?

602) Factor the polynomial.

$$6nmk^2 - 15mk + 14m^2nk^2 - 35m^2k$$

603) The ratio of the amount of money Luis has to the amount of money Chris has is 5:2. If Luis gives Chris $30, the two will then have equal amounts. Find the original amount that each one has.

Answer key

1)	12.1mph, 13.5mph	40)	2 hours	78)	9.7 hours
4)	$180	42)	9 pounds, 6 pounds	80)	5.3 miles per hour
6)	8	44)	20%	82)	82%
8)	0.8	46)	21 hours	84)	249 people
10)	0.9 gallons more	48)	3 hours	86)	$170.00
12)	0.8 pounds	50)	5 kilometers per hour	88)	20%
14)	430 trees, 258 trees	52)	34 candies, 17 candies, 25 candies	90)	12%
16)	600 bars, 650 bars	54)	60 candy bars	92)	$1,960.00, $980.00
18)	672 people	56)	4,200 toys	94)	550,000 square miles
20)	31.2 pounds	58)	63.35 miles per hour	96)	8.2 grams
22)	$3.50	60)	95 cents	98)	92%
24)	420 miles	62)	$193.20	100)	192 pages
26)	66 horses	64)	120.48 square feet	102)	185.8 pounds
28)	12 miles	66)	30%	104)	4.2 ounces, 6.3 ounces, 10.5 ounces
30)	55%	68)	95%	106)	300 grams, 40 grams, 60 grams
32)	0.1 kilometers per minute	70)	$1.40	108)	10 inches
34)	10 kilometers per hour	72)	13 minutes	110)	495 liters
36)	72.5 miles per hour	74)	12 square feet		
38)	in 2 hours	76)	8.5 hours		

112) 3.2 miles per hour, 12.8 miles per hour

114) 50 miles per hour, 55 miles per hour

116) $90.10

118) 4 books

120) 43 points

122) 150 books

124) 105,000 travelers

126) 10 trees

128) 70 shirts

130) 3 hours

132) 4 hours

134) 5.22, 20.88

136) 6.16, 18.48

138) 12 pounds, 6 pounds

140) 3 meters per second

142) 54

144) $7,364, $1,120

146) 2 centimeters per second

148) 15 miles per hour

149) 44 meters, 58 meters

151) by 30%

153) 50 miles per hour, 60 miles per hour

155) 45 miles per hour

157) by 25%

159) 5.5 hours

161) 6 pounds, 3 pounds

163) 6 kilometers per hour

165) 825 miles, 82.5 miles

167) 5.71 miles per hour

169) 420 miles, 70 miles

171) 10 kilometers per hour

173) $23, $31

175) 40 letters, 43 letters, 36 letters

177) 20 pounds, 60 pounds

179) 8 inches, 6 inches

181) 40 miles per hour

183) 3 bushels

185) 4.2 tones, 3.6 tones

187) 3 gallons per hour, 4.5 gallons per hour

189) 15 miles, 18 miles

191) 13 breeds, 13 breeds, 26 breeds, 16 breeds

193) 36, 162

195) 72 miles per hour, 18 miles

197) $7, $5, $9

199) 125%

201) by 12.5%

203) $819

205) 30 pounds

207) 20 miles

209) 4,000 pieces

211) 300 movies

213) 70 cents

215) 85 pumpkins

217) 70 feet

219) $.60

222) 100 calories

224) 500,000 immigrants

226) $1.85

228) 237 students, 79 students, 114 students

234) 200 people, 176 people

238) 60 miles

240) $70

242) 51 centimeters, 63 centimeters

244) 82 people, 162 people

246) 600 turkeys, 1,800 turkeys

248) $280

250) 25

252) 28 people

254) 41 kids, 3 oranges, 2 apples

256) 17 buses

258) 6 jars, 3 jars

261) 8 magazines, 17 books

263) $4

267) 12 coins

269) 12 days

271) 120 days

273) 50%, 40%, 10%

275) 101.25 cubic feet

277) 12.5 miles per hour, 60 miles

279) 2,016 cubic inches

281) 3.5 hours

283) 5 hours

285) 7:05AM

287) 5:50PM

289) $400

291) 6 miles per hour

293) 512 inches

295) 336 words per minute

299) 16.5 inches

301) 3,450 bushels

303) 661 square inches

305) 30 feet, 60 feet

315) at 4PM

317) $7, $35, $17

319) $32, $24, $9

327) $200

329) $50

331) 15 chairs

333) 30 cakes

335) 429 meters

339) $\frac{4}{15}$

341) $400

343) 20 pounds

345) 9 desks, 16 desks

347) 48 condominiums, 24 condominiums

351) 32 pounds

353) 1,750 years

355) $\frac{7}{16}$

357) $\frac{20}{7}$

359) 12 horses, 7 horses, 2 horses

361) $2,800, $350, $2,450

363) 30 coins, 45 coins

365) 40 people, 100 people

367) 128%, by 28%

369) 2.5 times

371) $870

373) $144

375) 8.2, 8.9, 9.6

377) $1.70

379) 108 kilograms

381) 85%

383) 60 questions

385) 252 kilometers

387) 20 centimeters

389) 390 calories

391) 8 hours

393) $31\frac{1}{4}$ kilometers

395) 36 kilograms

397) $10.45

399) 6.03 tons

401) 13.14 kilograms

403) 6.25%

405) 18 pounds, 16 pounds

407) $8

409) 8.2 inches, 9.2 inches, 10.1 inches

411) 0.8 pounds

413) 4.2 kilograms

415) 21 miles per hour

417) $17\frac{7}{9}$ miles

419) 14 hours

421) 6 days

423) $233.60, $116.80, $350.40

425) $28, $35, $50

427) 40.5 miles per hour

429) $1\frac{3}{4}$ pounds

431) $69.30

433) 24 jellybeans, 64 jellybeans

435) 12 workers, 14 workers

437) 9 pounds

439) 375 liters, 350 liters

441) 30, 45

443) 56 people

445) 270 marbles

447) $98

449) $270

453) 135, 42

455) 40 cars, 20 cars

457) 25 coats, 50 coats

459) 2.4 hours

461) 900 miles

463) $3.06

465) 24 days

467) 10 hours, 15 hours

469) In $1\frac{3}{7}$ days

471) 14 square miles

473) $45.05

475) $20\frac{1}{4}$ miles

477) $525

479) 98 pages

481) $3,000

483) 33,600 bricks

485) 2.2 hours

487) 104 miles, 65 miles

489) 5.5 hours

491) 12,800 cubic yards

493) $17,600

495) 21%

497) $1\frac{4}{5}$ tones

499) $1,330, $950

501) 180 gallons

503) 153, 81

505) $22.50, $30, $33

507) 42 miles

509) 6 days

511) 180 planes, 200 planes

513) $12, $16

515) 60 miles per hour, 75 miles per hour

517) 40 liters per minute, 120 liters per minute

519) 2 hours

521) 5 miles per hour

523) 15, 22, 29, 36

525) 53.1 miles per hour

527) 700 tickets

529) 2 hours

531) 5 hours

533) 54 miles per hour

535) 4 hours

537) 17.5 miles per hour, 20.5 miles per hour

539) 12 crayons, 6 crayons

541) 1,350 pounds, 225 pounds

543) 30, 90

545) 66 years, 33 years, 11 years

547) 18 pages

549) 320 students

551) 120 grams, 80 grams, 40 grams

553) 96 pages

555) $72

557) 25 meters, 7.5 meters

559) 16 pastries

561) 18 problems

563) 7 pillowcases

565) 25 pounds, 50 pounds

567) 0.3 pounds, 1.2 pounds

569) 9.6 meters

571) 62.5 miles per hour

573) $14,678.50

575) $9,800

577) 1.5 hours

579) 4 hours

581) 14 coins, 7 coins, 12 coins

583) 0.25 hours

585) 1.5, 1.5, 2.4

589) 0.35 kilograms, 1.05 kilograms

591) 1.1 pounds, 1.65 pounds

597) $1\frac{5}{7}$ hours

599) 95

601) 18 hours

603) $100, $40

0-595-32185-2